Differentiated Instruction Teacher Management System

Holt Social Studies

Africa

HOLT, RINEHART AND WINSTON
A Harcourt Education Company
Orlando • Austin • New York • San Diego • Toronto • London

[Acknowledgments and other credits to appear here.]

Printed in the United States of America

ISBN 0-03-078516-2

3 4 5 6 7 8 9 10 018 09 08 07

Contents

The *Differentiated Instruction Teacher Management System* is designed to assist you in planning and managing instruction for all students who are using *Holt Social Studies: Africa.* This book has four parts: a Pacing Guide, Section Lesson Plans, Lesson Plans for Differentiated Instruction, and the Teaching Guide and Answer Key for the *Interactive Reader and Study Guide.*

Pacing Guide

This part of the *Differentiated Instruction Teacher Management System* includes a pacing guide—broken down into a 9-week instructional segment and a 12-week instructional segment. The guide suggests materials to be taught and assessed during an average 45-minute class period.

Section Lesson Plans

These on-level lesson plans provide a basic form and planning guide for each section in the textbook. Each lesson plan includes the following content:

- Objectives
- Vocabulary
- Lesson Plan Organization

Lesson Plans for Differentiated Instruction

These lesson plans provide strategies and suggestions for teaching students with special needs. Differentiated strategies are included for:

- English-Language Instruction
- Special Education Instruction
- Advanced/Gifted and Talented Instruction

Teaching Guide and Answer Key for the Interactive Reader and Study Guide

This guide and accompanying answer key are provided to support the *Interactive Reader and Study Guide*, a student workbook that accompanies *Holt Social Studies: Africa.*

LESSON AND ACTIVITY	9-Week Course	12-Week Course
	(Number of 45-minute class periods)	
Progress Assessment: Diagnostic Test	1	1
Chapter: History of Ancient Egypt, 4500–500 BC	**6**	**8**
Section 1: Geography and Ancient Egypt	1	1
Section 2: The Old Kingdom	1	2
Close-up: Building the Pyramids		
Section 3: The Middle and New Kingdoms	1	2
Biography: Queen Hatshepsut		
Biography: Ramses the Great		
Section 4: Egyptian Achievements	2	2
Close-up: The Temple of Karnak		
Social Studies Skills: Analyzing Primary and Secondary Sources		
Chapter Review and Test	1	1
Chapter: History of Ancient Kush, 2300 BC–AD 350	**5**	**6**
Section 1: Kush and Egypt	2	2.5
Biography: Piankhi		
Section 2: Later Kush	2	2.5
Biography: Queen Shanakhdakheto		
Close-up: Rulers of Kush		
Social Studies Skills: Identifying Bias		
Chapter Review and Test	1	1
Chapter: History of West Africa, 500 BC–AD 1650	**5**	**7**
Section 1: The Empire of Ghana	1.5	2
Biography: Tunka Manin		
Geography and History: Crossing the Sahara		
Section 2: Mali and Songhai	1	2
Biography: Askia the Great		
Biography: Mansa Musa		
Section 3: Historical and Artistic Traditions	1.5	2
Connecting to the Arts: Music from Mali to Memphis		
Social Studies Skills: Making Decisions		
Chapter Review and Test	1	1

Chapter: North Africa	**5**	**7**
Section 1: Physical Geography	1	2
Close-up: The Sahara		
Section 2: History and Culture	1.5	2
Biography: Cleopatra		
Focus on Culture: The Berbers		
Social Studies Skills: Analyzing a Diagram		
Section 3: North Africa Today	1.5	2
Satellite View: The Nile River		
World Almanac Facts About Countries: Percentage of Egypt's Rural and Urban Population		
Chapter Review and Test	1	1

Chapter: West Africa	**5**	**7**
Section 1: Physical Geography	1	2
Section 2: History and Culture	1.5	2
Primary Source: I Speak of Freedom		
Connecting to the Arts: Masks		
Geography and History: The Atlantic Slave Trade		
Section 3: West Africa Today	1.5	2
World Almanac Facts About Countries: Nigeria's Cities		
Literature: The Years of Childhood		
Social Studies Skills: Analyzing a Precipitation Map		
Chapter Review and Test	1	1

Chapter: East Africa	**5**	**7**
Section 1: Physical Geography	1	1.5
Satellite View: Great Rift Valley		
Section 2: History and Culture	1.5	2
Focus on Culture: Swahili		
Section 3: East Africa Today	1.5	2.5
Close-up: The Serengeti Plain		
World Almanac Facts About Countries: Population Density of East Africa		
Social Studies Skills: Doing Fieldwork and Using Questionnaires		
Chapter Review and Test	1	1

Chapter: Central Africa	**6**	**8**
Section 1: Physical Geography *Satellite View:* Deforestation in the Congo *Case Study:* Mapping Central Africa's Forests	1.5	2
Section 2: History and Culture *World Almanac Facts About Countries:* Religions of Central Africa *Focus on Culture:* Foods of Central Africa	1.5	2
Section 3: Central Africa Today *World Almanac Facts About Countries:* Population of Kinshasa *Social Studies Skills:* Interpreting a Population Pyramid	2	3
Chapter Review and Test	1	1

Chapter: Southern Africa	**6**	**8**
Section 1: Physical Geography *Satellite View:* Namib Desert	1.5	2
Section 2: History and Culture *Focus on Culture:* Zulu Music	1.5	2
Section 3: Southern Africa Today *Biography:* Nelson Mandela *Connecting to Economics:* Tourism *World Almanac Facts About Countries:* Number of Visitors to Southern Africa *Social Studies Skills:* Evaluating a Website	2	3
Chapter Review and Test	1	1

Progress Assessment: Benchmark Test, Forms A and B	1	1

Additional Teacher Resources to Support **Differentiated Instruction**
Interactive Reader and Study Guide
Africa Transparencies
Chapter Visual Summary
Social Studies Skills Activity
Interactive Skills Tutor CD-ROM
Differentiated Instruction Modified Worksheets and Tests CD-ROM

Teacher ______________________ Date______________

History of Ancient Egypt

Lesson Plan

Section 1

Objectives *Students will learn that . . .*

1. Egypt was called the gift of the Nile because the Nile River was so important.
2. Civilization developed after people began farming along the Nile River.
3. Strong kings unified all of ancient Egypt.

Vocabulary Nile River, Upper Egypt, Lower Egypt, cataracts, delta, pharaoh, dynasty

PRETEACH	RESOURCES
___ **What You Will Learn . . .** (SE) Preview Main Ideas, Big Idea, and Key Terms. ___ **Key Terms and Places** (TE) Preteach or review the key terms from this section with students. ___ **If YOU lived there . . .** (SE) Have students read and discuss the section introduction and question.	___ Daily Bellringer Transparency: Section 1 ___ RF: Vocabulary Builder: Section 1 ___ Differentiated Instruction Modified Worksheets and Tests CD-ROM
DIRECT TEACH	**RESOURCES**
___ **Teach the Big Idea Activity** (TE) Students discuss the Main Idea questions and then write verses to a national anthem celebrating ancient Egyptian unification. ___ **Main Idea Questions** (TE) Discussion questions ___ **Map: Ancient Egypt** (SE) ___ **Collaborative Learning** (TE) A 3-D Map of Ancient Egypt ___ **Differentiating Instruction: English-Language Learners** (TE) Nile Civilization Diagram ___ **Critical Thinking: Analyzing Information** (TE) The Double Crown of Menes	___ Interactive Reader and Study guide: Section 1 ___ Map Zone Transparency: Ancient Egypt
REVIEW & ASSESS	**RESOURCES**
___ **Close** (TE) Call on volunteers to compose additional questions about the illustrations in this section. ___ **Section 1 Assessment** (SE)	___ Online Quiz Section 1 (keyword: SGE7 HP1) ___ PASS: Section 1 Quiz

Key: **SE** = Student Edition **TE** = Teacher's Edition **RF** = Resource File **PASS** = Progress Assessment Support System

Teacher ______________________________ Date______________________

History of Ancient Egypt — Lesson Plan

Section 2

Objectives *Students will learn that . . .*

1. Life in the Old Kingdom was influenced by pharaohs, roles in society, and trade.
2. Religion shaped Egyptian life.
3. The pyramids were built as tombs for Egypt's pharaohs.

Vocabulary Old Kingdom, nobles, afterlife, mummies, elite, pyramids, engineering

PRETEACH	RESOURCES
___ **What You Will Learn . . .** (SE) Preview Main Ideas, Big Idea, and Key Terms. ___ **Academic Vocabulary** (TE) Review with students the high-use academic terms in this section. ___ **If YOU lived there . . .** (SE) Have students read and discuss the section introduction and question.	___ Daily Bellringer Transparency: Section 2 ___ RF: Vocabulary Builder: Section 2 ___ Differentiated Instruction Modified Worksheets and Tests CD-ROM
DIRECT TEACH	**RESOURCES**
___ **Teach the Big Idea Activity** (TE) Students discuss the Main Idea questions and then design stamps to commemorate the anniversary of the Old Kingdom's beginning. ___ **Main Idea Questions** (TE) Discussion questions ___ **Critical Thinking: Evaluating** (TE) Being Pharaoh ___ **Differentiating Instruction: Advanced/Gifted and Talented** (TE) Egyptian Gods ___ **Collaborative Learning** (TE) Egyptian Game Show ___ **Social Studies Skill: Identifying Central Issues** (TE) Museum Exhibit ___ **Close-up: Building the Pyramids** (SE) ___ **Cross-Discipline Activity: Math** (TE) The Geometry of Pyramids ___ **Social Studies Skills: Developing Group Interaction Skills** (TE) Building a Pyramid	___ Interactive Reader and Study Guide: Section 2 ___ RF: Biography: Khufu
REVIEW & ASSESS	**RESOURCES**
___ **Close** (TE) Ask students to imagine what went through the workers' minds as they built the pyramids. ___ **Section 2 Assessment** (SE)	___ Online Quiz Section 2 (keyword: SGE7 HP1) ___ PASS: Section 2 Quiz

Key: **SE** = Student Edition **TE** = Teacher's Edition **RF** = Resource File **PASS** = Progress Assessment Support System

Teacher ______________________ Date ______________________

History of Ancient Egypt

Lesson Plan

Section 3

Objectives *Students will learn that . . .*

1. The Middle Kingdom was a period of stable government between periods of disorder.
2. The New Kindgom was the peak of Egyptian trade and military power, but its greatness did not last.
3. Work and daily life differed among Egypt's social classes.

Vocabulary Middle Kingdom, New Kingdom, Kush, trade routes

PRETEACH	RESOURCES
___ **What You Will Learn . . .** (SE) Preview Main Ideas, Big Idea, and Key Terms. ___ **Academic Vocabulary** (TE) Review with students the high-use academic term in this section. ___ **If YOU lived there . . .** (SE) Have students read and discuss the section introduction and question.	___ Daily Bellringer Transparency: Section 3 ___ RF: Vocabulary Builder: Section 3 ___ Differentiated Instruction Modified Worksheets and Tests CD-ROM
DIRECT TEACH	**RESOURCES**
___ **Teach the Big Idea Activity** (TE) Students discuss the Main Idea questions and then create time lines highlighting prominent rulers and their accomplishments. ___ **Main Idea Questions** (TE) Discussion questions ___ **Time Line: Periods of Egyptian History** (SE) ___ **Biography: Queen Hatshepsut** (SE) ___ **Critical Thinking: Interpreting Maps** (TE) Trade Routes of Ancient Egypt ___ **Map: Egyptian Trade, c. 1400 BC** (SE) ___ **Critical Thinking: Summarizing** (TE) Two New Kingdom Rulers ___ **Differentiating Instruction: Struggling Readers** (TE) Egyptian Society Comparison ___ **Differentiating Instruction: Advanced/Gifted and Talented** (TE) Egyptian Traders	___ Interactive Reader and Study Guide: Section 3 ___ Map Zone Transparency: Egyptian Trade, c. 1400 BC ___ RF: Geography and History: Trade on the Nile ___ RF: Biography: Nefertiti
REVIEW & ASSESS	**RESOURCES**
___ **Close** (TE) Point out that Egyptian children played versions of tug-of-war, leap-frog, and checkers. Ask how life in ancient Egypt was similar to and differed from life today. ___ **Section 3 Assessment** (SE)	___ Online Quiz Section 3 (keyword: SGE7 HP1) ___ PASS: Section 3 Quiz

Key: **SE** = Student Edition **TE** = Teacher's Edition **RF** = Resource File **PASS** = Progress Assessment Support System

Teacher ______________________ Date ______________

History of Ancient Egypt

Lesson Plan

Section 4

Objectives *Students will learn that . . .*

1. Egyptian writing used symbols called hieroglyphics.
2. Egypt's great temples were lavishly decorated.
3. Egyptian art filled tombs.

Vocabulary hieroglyphics, papyrus, Rosetta Stone, sphinxes, obelisk

PRETEACH	RESOURCES
___ **What You Will Learn . . .** (SE) Preview Main Ideas, Big Idea, and Key Terms. ___ **Key Terms and Places** (TE) Preteach or review the key terms from this section with students. ___ **If YOU lived there . . .** (SE) Have students read and discuss the section introduction and question.	___ Daily Bellringer Transparency: Section 4 ___ RF: Vocabulary Builder: Section 4 ___ Differentiated Instruction Modified Worksheets and Tests CD-ROM
DIRECT TEACH	**RESOURCES**
___ **Teach the Big Idea Activity** (TE) Students discuss the Main Idea questions and then write autobiographies, assuming the perspective of a contributor to Egyptian achievements. ___ **Main Idea Questions** (TE) Discussion questions ___ **Differentiating Instruction: English-Language Learners** (TE) Rosetta Stone Translation ___ **Close-up: The Temple of Karnak** (SE) ___ **Cross-Discipline Activity: Math** (TE) Measuring a Temple ___ **Cross-Discipline Activity: Arts and the Humanities** (TE) The Opet Festival ___ **Collaborative Learning** (TE) King Tut's Tomb ___ **Social Studies Skills: Analyzing Primary and Secondary Sources** (SE)	___ Interactive Reader and Study Guide: Section 4 ___ RF: Literature: *The Book of the Dead: The Papyrus of Ani* ___ RF: Primary Source: Hymn to the Nile ___ RF: Primary Source: Ancient Egyptian Tomb Paintings
REVIEW & ASSESS	**RESOURCES**
___ **Close** (TE) Have students describe the works of art shown on these pages. ___ **Section 4 Assessment** (SE)	___ Quick Facts Transparency: History of Ancient Egypt Visual Summary ___ Online Quiz Section 4 (keyword: SGE7 HP1) ___ PASS: Section 4 Quiz

Key: **SE** = Student Edition **TE** = Teacher's Edition **RF** = Resource File **PASS** = Progress Assessment Support System

Teacher ______________________________ Date______________

History of Ancient Kush

Lesson Plan

Section 1

Objectives *Students will learn that . . .*

1. Geography helped early Kush civilization develop in Nubia.
2. Egypt controlled Kush for about 450 years.
3. After winning its independence, Kush ruled Egypt and set up a new dynasty there.

Vocabulary Nubia, ebony, ivory

PRETEACH	RESOURCES
___ **What You Will Learn . . .** (SE) Preview Main Ideas, Big Idea, and Key Terms. ___ **Key Terms and Places** (TE) Preteach or review the key terms from this section with students. ___ **If YOU lived there . . .** (SE) Have students read and discuss the section introduction and question.	___ Daily Bellringer Transparency: Section 1 ___ RF: Vocabulary Builder: Section 1 ___ Differentiated Instruction Modified Worksheets and Tests CD-ROM
DIRECT TEACH	**RESOURCES**
___ **Teach the Big Idea Activity** (TE) Students discuss the Main Idea questions and then plan a billboard using their knowledge of Egypt's relations with ancient Kush. ___ **Main Idea Questions** (TE) Discussion questions ___ **Map: Ancient Kush** (SE) ___ **Differentiating Instruction: Advanced/Gifted and Talented** (TE) Elevation Profiles of Kush ___ **Critical Thinking: Analyzing Information** (TE) Charting Natural Resources ___ **Biography: Piankhi** (SE) ___ **Collaborative Learning** (TE) Victory Newspaper	___ Interactive Reader and Study Guide: Section 1 ___ RF: Geography and History: The Nile ___ Map Zone Transparency: Ancient Kush ___ RF: Biography: King Taharqa
REVIEW & ASSESS	**RESOURCES**
___ **Close** (TE) Call on volunteers to pose new questions about the images in this section and for other students to suggest answers. ___ **Section 1 Assessment** (SE)	___ Online Quiz Section 1 (keyword: SGE7 HP2) ___ PASS: Section 1 Quiz

Key: **SE** = Student Edition **TE** = Teacher's Edition **RF** = Resource File **PASS** = Progress Assessment Support System

Teacher ______________________ Date ______________________

History of Ancient Kush

Lesson Plan

Section 2

Objectives *Students will learn that . . .*

1. Kush's economy grew because of its iron industry and trade network.
2. Some elements of Kushite society and culture were borrowed from other cultures while others were unique to Kush.
3. The decline and defeat of Kush was caused by both internal and external factors.

Vocabulary Meroë, trade network, merchants, exports, imports

PRETEACH	RESOURCES
___ **What You Will Learn . . .** (SE) Preview Main Ideas, Big Idea, and Key Terms. ___ **Academic Vocabulary** (TE) Review with students the high-use academic term in this section. ___ **If YOU lived there . . .** (SE) Have students read and discuss the section introduction and question.	___ Daily Bellringer Transparency: Section 2 ___ RF: Vocabulary Builder: Section 2 ___ Differentiated Instruction Modified Worksheets and Tests CD-ROM
DIRECT TEACH	**RESOURCES**
___ **Teach the Big Idea Activity** (TE) Students discuss the Main Idea questions and then compose a magazine article on a specific facet of Kush civilization. ___ **Main Idea Questions** (TE) Discussion questions ___ **Close-up: Kush's Trade Network** (SE) ___ **Differentiating Instruction: Advanced/Gifted and Talented** (TE) Iron Making Skit ___ **Biography: Queen Shanakhdakheto** (SE) ___ **Differentiating Instruction: English-Language Learners** (TE) Kush Venn Diagram ___ **Close-up: Rulers of Kush** (SE) ___ **Differentiating Instruction: Special Needs Learners** (TE) Stone Carvings ___ **Social Studies Skills: Identifying Bias** (SE)	___ Interactive Reader and Study Guide: Section 2 ___ RF: Biography: Queen Amanirenas ___ RF: Primary Source: Herodotus, *The Histories* ___ RF: Literature: The Fall of Meroë
REVIEW & ASSESS	**RESOURCES**
___ **Close** (TE) Review the reasons that a large trading network was important to Kush. ___ **Section 2 Assessment** (SE)	___ Quick Facts Transparency: History of Ancient Kush Visual Summary ___ Online Quiz Section 2 (keyword: SGE7 HP2) ___ PASS: Section 2 Quiz

Key: **SE** = Student Edition **TE** = Teacher's Edition **RF** = Resource File **PASS** = Progress Assessment Support System

Teacher ____________________ Date____________

History of West Africa

Lesson Plan

Section 1

Objectives *Students will learn that . . .*

1. Ghana controlled trade and became wealthy.
2. Through its control of trade, Ghana built an empire.
3. Attacking invaders, overgrazing, and the loss of trade caused Ghana's decline.

Vocabulary silent barter

PRETEACH	RESOURCES
___ **What You Will Learn . . .** (SE) Preview Main Ideas, Big Idea, and Key Terms. ___ **Academic Vocabulary** (TE) Review with students the high-use academic term in this section. ___ **If YOU lived there . . .** (SE) Have students read and discuss the section introduction and question.	___ Daily Bellringer Transparency: Section 1 ___ RF: Vocabulary Builder: Section 1 ___ Differentiated Instruction Modified Worksheets and Tests CD-ROM
DIRECT TEACH	**RESOURCES**
___ **Teach the Big Idea Activity** (TE) Students discuss the Main Idea questions and then identify the overall main ideas of the section. ___ **Main Idea Questions** (TE) Discussion questions ___ **Differentiating Instruction: Advanced/Gifted and Talented** (TE) Creating a Web Page ___ **Map: Ghana Empire, c. 1050** (SE) ___ **Collaborative Learning** (TE) Using the Barter System ___ **Differentiating Instruction: Advanced/Gifted and Talented** (TE) Trade Tariffs ___ **Differentiating Instruction: Struggling Readers** (TE) Tribute System ___ **Biography: Tunka Manin** (SE) ___ **Critical Thinking: Sequencing** (TE) Illustrated Time Line	___ Interactive Reader and Study Guide: Section 1 ___ Map Zone Transparency: Ghana Empire, c. 1050
REVIEW & ASSESS	**RESOURCES**
___ **Close** (TE) Ask students to imagine what life was like in Ghana at the height of its empire about 800. Discuss the peace and prosperity of the empire. Ask students to imagine what life was like around 1200, when the empire began to fall apart. ___ **Section 1 Assessment** (SE)	___ Online Quiz Section 1 (keyword: SGE7 HP3) ___ PASS: Section 1 Quiz

Key: **SE** = Student Edition **TE** = Teacher's Edition **RF** = Resource File **PASS** = Progress Assessment Support System

Teacher ______________________ Date______________

History of West Africa

Lesson Plan

Section 2

Objectives *Students will learn that . . .*

1. The empire of Mali reached its height under the ruler Mansa Musa, but the empire fell to invaders in the 1400s.
2. The Songhai built a new Islamic empire in West Africa, conquering many of the lands that were once part of Mali.

Vocabulary Niger River, Timbuktu, mosque, Gao, Djenné

PRETEACH	RESOURCES
___ **What You Will Learn . . .** (SE) Preview Main Ideas, Big Idea, and Key Terms. ___ **Key Terms and Places** (TE) Preteach or review the key terms from this section with students. ___ **If YOU lived there . . .** (SE) Have students read and discuss the section introduction and question.	___ Daily Bellringer Transparency: Section 2 ___ RF: Vocabulary Builder: Section 2 ___ Differentiated Instruction Modified Worksheets and Tests CD-ROM
DIRECT TEACH	**RESOURCES**
___ **Teach the Big Idea Activity** (TE) Students discuss the Main Idea questions and then create a children's book that briefly outlines the history of Mali or Songhai. ___ **Main Idea Questions** (TE) Discussion questions ___ **Map: Mali and Songhai** (SE) ___ **Critical Thinking: Comparing and Contrasting** (TE) The Empires of West Africa ___ **Close-up: Timbuktu** (SE) ___ **Critical Thinking: Categorizing** (TE) The Rule of Mansa Musa ___ **Biography: Askia the Great** (SE) ___ **Differentiating Instruction: Struggling Readers** (TE) Accomplishments of Askia the Great	___ Interactive Reader and Study Guide: Section 2 ___ RF: Primary Source: Two Descriptions of Emperors of Mali ___ RF: Biography: Sundiata ___ Map Zone Transparency: Mali and Songhai ___ RF: Biography: Sunni Ali
REVIEW & ASSESS	**RESOURCES**
___ **Close** (TE) Have students create a time line of the key events in the history of the empires of Mali and Songhai. ___ **Section 2 Assessment** (SE)	___ Online Quiz Section 2 (keyword: SGE7 HP3) ___ PASS: Section 2 Quiz

Key: **SE** = Student Edition **TE** = Teacher's Edition **RF** = Resource File **PASS** = Progress Assessment Support System

Teacher ______________________________ Date ______________________

History of West Africa

Lesson Plan

Section 3

Objectives *Students will learn that . . .*

1. West Africans have preserved their history through storytelling and the written accounts of visitors.
2. Through art, music, and dance, West Africans have expressed their creativity and kept alive their cultural traditions.

Vocabulary oral history, griots, proverbs, kente

PRETEACH	RESOURCES
___ **What You Will Learn . . .** (SE) Preview Main Ideas, Big Idea, and Key Terms. ___ **Key Terms** (TE) Preteach or review the key terms from this section with students. ___ **If YOU lived there . . .** (SE) Have students read and discuss the section introduction and question.	___ Daily Bellringer Transparency: Section 3 ___ RF: Vocabulary Builder: Section 3 ___ Differentiated Instruction Modified Worksheets and Tests CD-ROM
DIRECT TEACH	**RESOURCES**
___ **Teach the Big Idea Activity** (TE) Students discuss the Main Idea questions and then create a chart outlining the historical and artistic traditions of West Africa. ___ **Main Idea Questions** (TE) Discussion questions ___ **Differentiating Instruction: English-Language Learners** (TE) Oral Histories ___ **Connecting to the Arts: Music from Mali to Memphis** (SE) ___ **Critical Thinking: Analyzing Visuals** (TE) Art Appreciation ___ **Social Studies Skills: Making Decisions** (TE)	___ Interactive Reader and Study Guide: Section 3 ___ RF: Biography: Leo Africanus
REVIEW & ASSESS	**RESOURCES**
___ **Close** (TE) Have students discuss the elements of historical and artistic traditions that have had a lasting effect on West Africa. Ask students to identify elements that still exist today. ___ **Section 3 Assessment** (SE)	___ Quick Facts Transparency: History of West Africa Visual Summary ___ Online Quiz Section 3 (keyword: SGE7 HP3) ___ PASS: Section 3 Quiz

Key: **SE** = Student Edition **TE** = Teacher's Edition **RF** = Resource File **PASS** = Progress Assessment Support System

North Africa

Lesson Plan

Section 1

Objectives *Students will learn that . . .*

1. Major physical features of North Africa include the Nile River, the Sahara, and the Atlas Mountains.
2. The climate of North Africa is hot and dry, and water is the region's most important resource.

Vocabulary Sahara, Nile River, silt, Suez Canal, oasis, Atlas Mountains

PRETEACH	RESOURCES
___ **What You Will Learn . . .** (SE) Preview Main Ideas, Big Idea, and Key Terms. ___ **Academic Vocabulary** (TE) Review with students the high-use academic term in this section. ___ **If YOU lived there . . .** (SE) Have students read and discuss the section introduction and question.	___ Daily Bellringer Transparency: Section 1 ___ RF: Vocabulary Builder: Section 1 ___ Differentiated Instruction Modified Worksheets and Tests CD-ROM
DIRECT TEACH	**RESOURCES**
___ **Teach the Big Idea Activity** (TE) Students discuss the Main Idea questions and list the reasons why North Africa is a dry region. ___ **Main Idea Questions** (TE) Discussion questions ___ **Map: North Africa: Physical** (SE) ___ **Critical Thinking: Interpreting Maps** (TE) Important Physical Features of North Africa ___ **Close-up: A Sahara Oasis** (SE) ___ **Critical Thinking: Analyzing Visuals** (TE) Analyzing an Illustration	___ Interactive Reader and Study Guide: Section 1 ___ Map Zone Transparency: North Africa: Physical ___ RF: Geography for Life: Desertification
REVIEW & ASSESS	**RESOURCES**
___ **Close** (TE) Ask students to summarize the ways in which the Sahara and the Nile River influence life in North Africa. ___ **Section 1 Assessment** (SE)	___ Online Quiz Section 1 (keyword: SGE7 HP4) ___ PASS: Section 1 Quiz

Key: **SE** = Student Edition **TE** = Teacher's Edition **RF** = Resource File **PASS** = Progress Assessment Support System

Teacher ______________________ Date______________

North Africa Lesson Plan

Section 2

Objectives *Students will learn that . . .*

1. North Africa's history includes ancient Egyptian civilization.
2. Islam influences the cultures of North Africa and most people speak Arabic.

Vocabulary Alexandria, Berbers

PRETEACH	RESOURCES
___ **What You Will Learn . . .** (SE) Preview Main Ideas, Big Idea, and Key Terms. ___ **Key Terms and Places** (TE) Preteach or review the key terms from this section with students. ___ **If YOU lived there . . .** (SE) Have students read and discuss the section introduction and question.	___ Daily Bellringer Transparency: Section 2 ___ RF: Vocabulary Builder: Section 2 ___ Differentiated Instruction Modified Worksheets and Tests CD-ROM
DIRECT TEACH	**RESOURCES**
___ **Teach the Big Idea Activity** (TE) Students discuss the Main Idea questions and then identify facts about North Africa's history and culture. ___ **Main Idea Questions** (TE) Discussion questions ___ **Biography: Cleopatra** (SE) ___ **Differentiating Instruction: Struggling Readers** (TE) Impact of Arab Rule on North Africa Time Line ___ **Differentiating Instruction: Special Needs Learners** (TE) King Tut's Burial Chamber ___ **Differentiating Instruction: Advanced/Gifted and Talented** (TE) Research Ancient Egyptian Burial Customs ___ **Collaborative Learning** (TE) North African Cultures ___ **Focus On Culture: The Berbers** (SE) ___ **Differentiating Instruction: English-Language Learners** (TE) Culture and Food ___ **Social Studies Skills: Analyzing a Diagram** (SE)	___ Interactive Reader and Study Guide: Section 2 ___ RF: Literature: "The Shipwrecked Sailor" ___ RF: Critical Thinking: The Geometry of Ancient Egypt's Pyramids ___ RF: Primary Source: *Darkest Africa* ___ RF: Biography: Anwar al-Sadat ___ RF: Biography: Assia Djebar
REVIEW & ASSESS	**RESOURCES**
___ **Close** (TE) Ask students to identify examples of Arab influences on North African cultures. ___ **Section 2 Assessment** (SE)	___ Online Quiz Section 2 (keyword: SGE7 HP4) ___ PASS: Section 2 Quiz

Key: **SE** = Student Edition **TE** = Teacher's Edition **RF** = Resource File **PASS** = Progress Assessment Support System

Teacher ______________________________ Date______________

North Africa

Lesson Plan

Section 3

Objectives *Students will learn that . . .*

1. Many of Egypt's people are farmers and live along the Nile River.
2. People in the other countries of North Africa are mostly pastoral nomads or farmers, and oil is an important resource in the region.

Vocabulary Cairo, Maghreb, souks, free port, dictator

PRETEACH	RESOURCES
___ **What You Will Learn . . .** (SE) Preview Main Ideas, Big Idea, and Key Terms. ___ **Key Terms and Places** (TE) Preteach or review the key terms from this section with students. ___ **If YOU lived there . . .** (SE) Have students read and discuss the section introduction and question.	___ Daily Bellringer Transparency: Section 3 ___ RF: Vocabulary Builder: Section 3 ___ Differentiated Instruction Modified Worksheets and Tests CD-ROM
DIRECT TEACH	**RESOURCES**
___ **Teach the Big Idea Activity** (TE) Students discuss the Main Idea questions and then create a chart of the major cities, economies, and challenges facing North Africa today. ___ **Main Idea Questions** (TE) Discussion questions ___ **Map: Egypt: Population** (SE) ___ **The World Almanac: Facts about Countries: Egypt's Population, 2003** (SE) ___ **Differentiating Instruction: Struggling Readers** (TE) Population of Egypt ___ **Satellite View: The Nile River** (SE) ___ **Collaborative Learning** (TE) Evaluating Egypt's Challenges ___ **Map: North Africa: Agriculture** (SE) ___ **Collaborative Learning** (TE) Interpreting an Agricultural Map of North Africa ___ **Differentiating Instruction: Special Needs Learners** (TE) Reviewing the Maghreb	___ Interactive Reader and Study Guide: Section 3 ___ Map Zone Transparency: Egypt: Population ___ Satellite View Transparency: The Nile River ___ RF: Geography and History: Major Cities of North Africa ___ Map Zone Transparency: North Africa: Agriculture
REVIEW & ASSESS	**RESOURCES**
___ **Close** (TE) Have students describe the ways in which life in Egypt is similar to and different from life in the other North African countries. ___ **Section 3 Assessment** (SE)	___ Quick Facts Transparency: North Africa Visual Summary ___ Online Quiz Section 3 (keyword: SGE7 HP4) ___ PASS: Section 3 Quiz

Key: **SE** = Student Edition **TE** = Teacher's Edition **RF** = Resource File **PASS** = Progress Assessment Support System

Teacher ______________________________ Date______________

West Africa

Lesson Plan

Section 1

Objectives *Students will learn that . . .*

1. West Africa's key physical features include plains and the Niger River.
2. West Africa has distinct climate and vegetation zones that go from arid in the north to tropical in the south.
3. West Africa has good agricultural and mineral resources that may one day help the economies in the region.

Vocabulary Niger River, zonal, Sahel, desertification, savanna

PRETEACH	RESOURCES
___ **What You Will Learn . . .** (SE) Preview Main Ideas, Big Idea, and Key Terms. ___ **Key Terms and Places** (TE) Preteach or review the key terms from this section with students. ___ **If YOU lived there . . .** (SE) Have students read and discuss the section introduction and question.	___ Daily Bellringer Transparency: Section 1 ___ RF: Vocabulary Builder: Section 1 ___ Differentiated Instruction Modified Worksheets and Tests CD-ROM
DIRECT TEACH	**RESOURCES**
___ **Teach the Big Idea Activity** (TE) Students discuss the Main Idea questions and then write details about West Africa's physical features, climate and vegetation, and resources. ___ **Main Idea Questions** (TE) Discussion questions ___ **Map: West Africa: Physical** (SE) ___ **Differentiating Instruction: Advanced/Gifted and Talented** (TE) Major River Deltas ___ **Differentiating Instruction: English-Language Learners** (TE) The Greek Symbol for Delta ___ **Map: West Africa: Climate** (SE) ___ **Cross-Discipline Activity: Science** (TE) The Dust Bowl	___ Interactive Reader and Study Guide: Section 1 ___ Map Zone Transparency: West Africa: Physical ___ RF: Critical Thinking: Farming in Niger and Nigeria ___ Map Zone Transparency: West Africa: Climate
REVIEW & ASSESS	**RESOURCES**
___ **Close** (TE) Lead a class discussion about the physical features, climates and vegetation, and resources of West Africa. ___ **Section 1 Assessment** (SE)	___ Online Quiz Section 1 (keyword: SGE7 HP5) ___ PASS: Section 1 Quiz

Key: **SE** = Student Edition **TE** = Teacher's Edition **RF** = Resource File **PASS** = Progress Assessment Support System

Teacher ______________________ Date______________

West Africa Lesson Plan

Section 2

Objectives *Students will learn that . . .*

1. In West Africa's history, trade made great kingdoms rich, but this greatness declined as Europeans began to control trade routes.
2. The culture of West Africa includes many different ethnic groups, languages, religions, and housing styles.

Vocabulary Timbuktu, animism, extended family

PRETEACH	RESOURCES
___ **What You Will Learn . . .** (SE) Preview Main Ideas, Big Idea, and Key Terms. ___ **Key Terms and Places** (TE) Preteach or review the key terms from this section with students. ___ **If YOU lived there . . .** (SE) Have students read and discuss the section introduction and question.	___ Daily Bellringer Transparency: Section 2 ___ RF: Vocabulary Builder: Section 2 ___ Differentiated Instruction Modified Worksheets and Tests CD-ROM
DIRECT TEACH	**RESOURCES**
___ **Teach the Big Idea Activity** (TE) Students discuss the Main Idea questions and then discuss the connections between history and culture. ___ **Main Idea Questions** (TE) Discussion questions ___ **Primary Source: I Speak of Freedom** (SE) ___ **Differentiating Instruction: Advanced/Gifted and Talented** (TE) Research the Kingdoms of Ghana, Mali, and Songhai ___ **Differentiating Instruction: Struggling Readers** (TE) European Colonies in West Africa ___ **Connecting to the Arts: Masks** (SE) ___ **Critical Thinking Activity: Drawing Inferences** (TE) Extended Families	___ Interactive Reader and Study Guide: Section 2 ___ RF: Biography: Queen Amina of Zaria ___ RF: Geography for Life: Patterns of West African Migration ___ RF: Primary Source: African Wood Mask
REVIEW & ASSESS	**RESOURCES**
___ **Close** (TE) Briefly review the early kingdoms of West Africa, the slave trade, European colonization and independence, and the people, languages, religions, clothing, and homes of the region's people. ___ **Section 2 Assessment** (SE)	___ Online Quiz Section 2 (keyword: SGE7 HP5) ___ PASS: Section 2 Quiz

Key: **SE** = Student Edition **TE** = Teacher's Edition **RF** = Resource File **PASS** = Progress Assessment Support System

Teacher ____________________ Date ____________

West Africa

Lesson Plan

Section 3

Objectives *Students will learn that . . .*

1. Nigeria has many different ethnic groups, an oil-based economy, and one of the world's largest cities.
2. Most coastal countries of West Africa have struggling economies and weak or unstable governments.
3. Lack of resources in the Sahel countries is a main challenge to economic development.

Vocabulary secede, Lagos, famine

PRETEACH	RESOURCES
___ **What You Will Learn . . .** (SE) Preview Main Ideas, Big Idea, and Key Terms. ___ **Academic Vocabulary** (TE) Review with students the high-use academic term in this section. ___ **If YOU lived there . . .** (SE) Have students read and discuss the section introduction and question.	___ Daily Bellringer Transparency: Section 3 ___ RF: Vocabulary Builder: Section 3 ___ Differentiated Instruction Modified Worksheets and Tests CD-ROM
DIRECT TEACH	**RESOURCES**
___ **Teach the Big Idea Activity** (TE) Students discuss the Main Idea questions and then create a chart on West Africa today. ___ **Main Idea Questions** (TE) Discussion questions ___ **World Almanac: Facts About the World: Africa's Largest Cities** (SE) ___ **Collaborative Learning** (TE) Ethnic Groups in Nigeria ___ **Map: West Africa: Land Use and Resources** (SE) ___ **Differentiating Instruction: Advanced/Gifted and Talented** (TE) West African Griots ___ **Cross-Discipline Activity: Math** (TE) Comparing Income through a Graph ___ **Social Studies Skills: Analyzing a Precipitation Map** (SE)	___ Interactive Reader and Study Guide: Section 3 ___ RF: Literature: "Knowing Robs Us" ___ RF: Biography: Kwame Nkrumah ___ Map Zone Transparency: West Africa: Land Use and Resources
REVIEW & ASSESS	**RESOURCES**
___ **Close** (TE) Briefly review the current conditions of Nigeria and the coastal and Sahel countries. ___ **Section 3 Assessment** (SE)	___ Quick Facts Transparency: West Africa Visual Summary ___ Online Quiz Section 3 (keyword: SGE7 HP5) ___ PASS: Section 3 Quiz

Key: **SE** = Student Edition **TE** = Teacher's Edition **RF** = Resource File **PASS** = Progress Assessment Support System

Teacher ______________________________ Date______________

East Africa

Lesson Plan

Section 1

Objectives *Students will learn that . . .*

1. East Africa's physical features range from rift valleys to plains.
2. East Africa's climate is influenced by its location and elevation, and the region's vegetation includes savannas and forests.

Vocabulary rift valley, Great Rift Valley, Mount Kilimanjaro, Serengeti Plain, Lake Victoria, drought

PRETEACH	RESOURCES
___ **What You Will Learn . . .** (SE) Preview Main Ideas, Big Idea, and Key Terms. ___ **Key Terms and Places** (TE) Preteach or review the key terms from this section with students. ___ **If YOU lived there . . .** (SE) Have students read and discuss the section introduction and question.	___ Daily Bellringer Transparency: Section 1 ___ RF: Vocabulary Builder: Section 1 ___ Differentiated Instruction Modified Worksheets and Tests CD-ROM
DIRECT TEACH	**RESOURCES**
___ **Teach the Big Idea Activity** (TE) Students discuss the Main Idea questions and then create an itinerary for a visit to East Africa. ___ **Main Idea Questions** (TE) Discussion questions ___ **Map: East Africa: Physical** (SE) ___ **Differentiating Instruction: Struggling Readers** (TE) Identify Map Elements ___ **Differentiating Instruction: Advanced/Gifted and Talented** (TE) East African Board Game ___ **Satellite View: Great Rift Valley** (SE) ___ **Critical Thinking: Analyzing Information** (TE) Creating a Chronology	___ Interactive Reader and Study Guide: Section 1 ___ Map Zone Transparency: East Africa: Physical ___ RF: Geography for Life: The Disturbed Ecosystem of Lake Victoria ___ Map Zone Transparency: Great Rift Valley Satellite View
REVIEW & ASSESS	**RESOURCES**
___ **Close** (TE) Use the section's vocabulary to review important information about the physical geography of East Africa. ___ **Section 1 Assessment** (SE)	___ Online Quiz Section 1 (keyword: SGE7 HP6) ___ PASS: Section 1 Quiz

Key: **SE** = Student Edition **TE** = Teacher's Edition **RF** = Resource File **PASS** = Progress Assessment Support System

Teacher ____________________ Date ____________________

East Africa **Lesson Plan**

Section 2

Objectives *Students will learn that . . .*

1. The history of East Africa is one of religion, trade, and European influence.
2. East Africans speak many different languages and practice several different religions.

Vocabulary Nubia, Zanzibar, imperialism

PRETEACH	RESOURCES
___ **What You Will Learn . . .** (SE) Preview Main Ideas, Big Idea, and Key Terms. ___ **Key Terms and Places** (TE) Preteach or review the key terms from this section with students. ___ **If YOU lived there . . .** (SE) Have students read and discuss the section introduction and question.	___ Daily Bellringer Transparency: Section 2 ___ RF: Vocabulary Builder: Section 2 ___ Differentiated Instruction Modified Worksheets and Tests CD-ROM
DIRECT TEACH	**RESOURCES**
___ **Teach the Big Idea Activity** (TE) Students discuss the Main Idea questions and then create a chart on East Africa's history and culture. ___ **Main Idea Questions** (TE) Discussion questions ___ **Critical Thinking: Analyzing Information** (TE) Mapping Trade Patterns ___ **Focus on Culture: The Swahili** (SE) ___ **Cross-Discipline Activity: Literature** (TE) Reading About East Africa ___ **Map: Official Languages of East Africa** (SE)	___ Interactive Reader and Study Guide: Section 2 ___ RF: Geography and History: East African Slave Trade ___ RF: Literature: *Kiswahili* (The Swahili Language) ___ RF: Critical Thinking: African Proverbs ___ Map Zone Transparency: Official Languages of East Africa
REVIEW & ASSESS	**RESOURCES**
___ **Close** (TE) Have students create a time line of the events described in this section. ___ **Section 2 Assessment** (SE)	___ Online Quiz Section 2 (keyword: SGE7 HP6) ___ PASS: Section 2 Quiz

Key: **SE** = Student Edition **TE** = Teacher's Edition **RF** = Resource File **PASS** = Progress Assessment Support System

Teacher ______________________________ Date______________

East Africa

Lesson Plan

Section 3

Objectives *Students will learn that . . .*

1. National parks are a major source of income for Tanzania and Kenya.
2. Rwanda and Burundi are densely populated rural countries with a history of ethnic conflict.
3. Both Sudan and Uganda have economies based on agriculture, but Sudan has suffered from years of war.
4. The countries of the Horn of Africa are among the poorest in the world.

Vocabulary safari, geothermal energy, genocide, Darfur, Mogadishu

PRETEACH	RESOURCES
___ **What You Will Learn . . .** (SE) Preview Main Ideas, Big Idea, and Key Terms. ___ **Academic Vocabulary** (TE) Review with students the high-use academic term in this section. ___ **If YOU lived there . . .** (SE) Have students read and discuss the section introduction and question.	___ Daily Bellringer Transparency: Section 3 ___ RF: Vocabulary Builder: Section 3 ___ Differentiated Instruction Modified Worksheets and Tests CD-ROM
DIRECT TEACH	**RESOURCES**
___ **Teach the Big Idea Activity** (TE) Students discuss the Main Idea questions and then create a chart on East Africa today. ___ **Main Idea Questions** (TE) Discussion questions ___ **Close-up: Serengeti National Park** (SE) ___ **Differentiating Instruction: Special Needs Learners** (TE) Safari in the Serengeti; **English-Language Learners** (TE) Describing the Visuals ___ **Map: East Africa: Population** (SE) ___ **World Almanac: Facts About Countries: Population Density in East Africa** (SE) ___ **Cross-Discipline Activity: Math** (TE) Creating a Cartogram ___ **Collaborative Activity** (TE) Time Line of Events ___ **Differentiating Instruction: Struggling Readers** (TE) Cause-and-Effect Chart ___ **Social Studies Skills: Doing Fieldwork and Using Questionnaries** (SE)	___ Interactive Reader and Study Guide: Section 3 ___ RF: Biography: Florence Wambugu ___ RF: Primary Source: Tinga Tinga Style of Art ___ Map Zone Transparency: East Africa: Population ___ RF: Biography: Paul Rusesabagina ___ RF: Interdisciplinary Project: Famine: An Ever-Present Threat
REVIEW & ASSESS	**RESOURCES**
___ **Close** (TE) Ask students what the countries in East Africa have in common and how they are different. ___ **Section 3 Assessment** (SE)	___ Quick Facts Transparency: East Africa Visual Summary ___ Online Quiz Section 3 (keyword: SGE7 HP6) ___ PASS: Section 3 Quiz

Key: **SE** = Student Edition **TE** = Teacher's Edition **RF** = Resource File **PASS** = Progress Assessment Support System

Teacher ______________________________ Date______________________

Central Africa — Lesson Plan

Section 1

Objectives *Students will learn that . . .*

1. Central Africa's major physical features include the Congo Basin and plateaus surrounding the basin.
2. Central Africa has a humid tropical climate and dense forest vegetation.
3. Central Africa's resources include forest products and valuable minerals such as diamonds and copper.

Vocabulary Congo Basin, basin, Congo River, Zambezi River, periodic market, copper belt

PRETEACH	RESOURCES
___ **What You Will Learn . . .** (SE) Preview Main Ideas, Big Idea, and Key Terms. ___ **Key Terms and Places** (TE) Preteach or review the key terms from this section with students. ___ **If YOU lived there . . .** (SE) Have students read and discuss the section introduction and question.	___ Daily Bellringer Transparency: Section 1 ___ RF: Vocabulary Builder: Section 1 ___ Differentiated Instruction Modified Worksheets and Tests CD-ROM
DIRECT TEACH	**RESOURCES**
___ **Teach the Big Idea Activity** (TE) Students discuss the Main Idea questions and then locate Central African countries on a map. ___ **Main Idea Questions** (TE) Discussion questions ___ **Map: Central Africa: Physical** (SE) ___ **Collaborative Learning** (TE) Managing the Congo River ___ **Map: Central Africa's National Parks** (SE) ___ **Differentiating Instruction: Special Needs Learners** (TE) Low Latitudes and Tropical Climate	___ Interactive Reader and Study Guide: Section 1 ___ RF: Geography and History: Rivers in the Democratic Republic of the Congo ___ Map Zone Transparency: Central Africa: Physical ___ Map Zone Transparency: Central Africa's National Parks
REVIEW & ASSESS	**RESOURCES**
___ **Close** (TE) Ask students to describe Central Africa's landforms, vegetation, climates, and rivers. ___ **Section 1 Assessment** (SE)	___ Online Quiz Section 1 (keyword: SGE7 HP7) ___ PASS: Section 1 Quiz

Key: **SE** = Student Edition **TE** = Teacher's Edition **RF** = Resource File **PASS** = Progress Assessment Support System

Teacher ______________________________ Date______________________

Central Africa

Lesson Plan

Section 2

Objectives *Students will learn that . . .*

1. Great African kingdoms and European colonizers have influenced the history of Central Africa.
2. The culture of Central Africa includes many ethnic groups and languages, but it has also been influenced by European colonization.

Vocabulary Kongo Kingdom, dialects

PRETEACH	RESOURCES
___ **What You Will Learn . . .** (SE) Preview Main Ideas, Big Idea, and Key Terms. ___ **Key Terms and Places** (TE) Preteach or review the key terms from this section with students. ___ **If YOU lived there . . .** (SE) Have students read and discuss the section introduction and question.	___ Daily Bellringer Transparency: Section 2 ___ RF: Vocabulary Builder: Section 2 ___ Differentiated Instruction Modified Worksheets and Tests CD-ROM
DIRECT TEACH	**RESOURCES**
___ **Teach the Big Idea Activity** (TE) Students discuss the Main Idea questions and then take notes on the key events, government, and culture of the region. ___ **Main Idea Questions** (TE) Discussion questions ___ **Differentiating Instruction: Special Needs Learners** (TE) Cause-and-Effect Graphic Organizer ___ **World Almanac: Facts about the World: Major Religions of Central Africa** (SE) ___ **Critical Thinking: Drawing Conclusions** (TE) Thinking About Traditions and Culture ___ **Connecting to History: Bantu Languages** (SE)	___ Interactive Reader and Study Guide: Section 2 ___ RF: Literature: "Western Civilization" ___ RF: Primary Source: The Malawi Constitution
REVIEW & ASSESS	**RESOURCES**
___ **Close** (TE) Have students identify three major periods in Central Africa's history and explain how colonial rule influenced the cultures of the region. ___ **Section 2 Assessment** (SE)	___ Online Quiz Section 2 (keyword: SGE7 HP7) ___ PASS: Section 2 Quiz

Key: **SE** = Student Edition **TE** = Teacher's Edition **RF** = Resource File **PASS** = Progress Assessment Support System

Teacher ______________________ Date ______________

Central Africa Lesson Plan

Section 3

Objectives *Students will learn that . . .*

1. The countries of Central Africa are mostly poor, and many are trying to recover from years of civil war.
2. Challenges to peace, health, and the environment slow economic development in Central Africa.

Vocabulary Kinshasa, inflation, malaria, malnutrition

PRETEACH	RESOURCES
___ **What You Will Learn . . .** (SE) Preview Main Ideas, Big Idea, and Key Terms. ___ **Academic Vocabulary** (TE) Review with students the high-use academic term in this section. ___ **If YOU lived there . . .** (SE) Have students read and discuss the section introduction and question.	___ Daily Bellringer Transparency: Section 3 ___ RF: Vocabulary Builder: Section 3 ___ Differentiated Instruction Modified Worksheets and Tests CD-ROM
DIRECT TEACH	**RESOURCES**
___ **Teach the Big Idea Activity** (TE) Students discuss the Main Idea questions and then identify challenges or problems facing Central Africa today. ___ **Main Idea Questions** (TE) Discussion questions ___ **World Almanac: Facts about Countries: Kinshasa's Growing Population** (SE) ___ **Differentiating Instruction: English-Language Learners** (TE) Stability and Instability in Central Africa ___ **Cross-Discipline Activity: Government** (TE) A Congo River Association ___ **Collaborative Learning** (TE) Geography Bowl ___ **Map: Malaria in Central Africa** (SE) ___ **Differentiating Instruction: Special Needs Learners** (TE) Malaria Posters ___ **Differentiating Instruction: Advanced/Gifted and Talented** (TE) Logging in Central Africa Debate ___ **Social Studies Skills: Interpreting a Population Pyramid** (SE)	___ Interactive Reader and Study Guide: Section 3 ___ RF: Biography: Patrice Lumumba ___ RF: Biography: Omar Bongo ___ RF: Critical Thinking: The Ebola Virus and Medical Geography ___ Map Zone Transparency: Malaria in Central Africa
REVIEW & ASSESS	**RESOURCES**
___ **Close** (TE) Have students describe the similar challenges faced by the governments and economies of Central Africa. ___ **Section 3 Assessment** (SE)	___ Quick Facts Transparency: Central Africa Visual Summary ___ Online Quiz Section 3 (keyword: SGE7 HP7) ___ PASS: Section 3 Quiz

Key: **SE** = Student Edition **TE** = Teacher's Edition **RF** = Resource File **PASS** = Progress Assessment Support System

Teacher ____________________ Date____________

Southern Africa Lesson Plan

Section 1

Objectives *Students will learn that . . .*

1. Southern Africa's main physical feature is a large plateau with plains, rivers, and mountains.
2. The climate and vegetation of Southern Africa is mostly savanna and desert.
3. Southern Africa has valuable mineral resources.

Vocabulary escarpment, veld, Namib Desert, pans

PRETEACH	RESOURCES
___ **What You Will Learn . . .** (SE) Preview Main Ideas, Big Idea, and Key Terms. ___ **Key Terms and Places** (TE) Preteach or review the key terms from this section with students. ___ **If YOU lived there . . .** (SE) Have students read and discuss the section introduction and question.	___ Daily Bellringer Transparency: Section 1 ___ RF: Vocabulary Builder: Section 1 ___ Differentiated Instruction Modified Worksheets and Tests CD-ROM
DIRECT TEACH	**RESOURCES**
___ **Teach the Big Idea Activity** (TE) Students discuss the Main Idea questions and then write facts about the countries of Southern Africa. ___ **Main Idea Questions** (TE) Discussion questions ___ **Map: Southern Africa: Physical** (SE) ___ **Differentiating Instruction: Struggling Readers** (TE) Map Labels ___ **Differentiating Instruction: Advanced/Gifted and Talented** (TE) Major Rivers of Southern Africa ___ **Satellite View: Namib Desert** (SE) ___ **Collaborative Learning** (TE) Mapping Climates ___ **Map: Southern Africa: Vegetation** (SE)	___ Interactive Reader and Study Guide: Section 1 ___ RF: Geography and History: Rivers of Southern Africa ___ Map Zone Transparency: Southern Africa: Physical ___ Africa Transparency: Satellite View: Namib Desert ___ Map Zone Transparency: Southern Africa: Vegetation
REVIEW & ASSESS	**RESOURCES**
___ **Close** (TE) Discuss with students the diverse physical features, climate, vegetation, and natural resources of Southern Africa. ___ **Section 1 Assessment** (SE)	___ Online Quiz Section 1 (keyword: SGE7 HP8) ___ PASS: Section 1 Quiz

Key: **SE** = Student Edition **TE** = Teacher's Edition **RF** = Resource File **PASS** = Progress Assessment Support System

Teacher ______________________ Date______________

Southern Africa Lesson Plan

Section 2

Objectives *Students will learn that . . .*

1. Southern Africa's history began with hunter-gatherers, followed by great empires and European settlements.
2. The cultures of Southern Africa are rich in different languages, religions, customs, and art.

Vocabulary Great Zimbabwe, Cape of Good Hope, Afrikaners, Boers, apartheid, township

PRETEACH	RESOURCES
___ **What You Will Learn . . .** (SE) Preview Main Ideas, Big Idea, and Key Terms. ___ **Key Terms and Places** (TE) Preteach or review the key terms from this section with students. ___ **If YOU lived there . . .** (SE) Have students read and discuss the section introduction and question.	___ Daily Bellringer Transparency: Section 2 ___ RF: Vocabulary Builder: Section 2 ___ Differentiated Instruction Modified Worksheets and Tests CD-ROM
DIRECT TEACH	**RESOURCES**
___ **Teach the Big Idea Activity** (TE) Students discuss the Main Idea questions and then create flowcharts on Southern African history. ___ **Main Idea Questions** (TE) Discussion questions ___ **Differentiating Instruction: Struggling Readers** (TE) Early History of Southern Africa ___ **Critical Thinking: Cause and Effect** (TE) The Effects of European Settlement ___ **Focus on Culture: Music of South Africa** (SE) ___ **Collaborative Learning** (TE) History Detectives ___ **Critical Thinking: Making Generalizations** (TE) Cultures of Southern Africa	___ Interactive Reader and Study Guide: Section 2 ___ RF: Literature: "Dream Life and Real Life, A Little African Story"
REVIEW & ASSESS	**RESOURCES**
___ **Close** (TE) Briefly review the history and culture of Southern Africa, asking volunteers to share a generalization they can make based on their reading. ___ **Section 2 Assessment** (SE)	___ Online Quiz Section 2 (keyword: SGE7 HP8) ___ PASS: Section 2 Quiz

Key: **SE** = Student Edition **TE** = Teacher's Edition **RF** = Resource File **PASS** = Progress Assessment Support System

Southern Africa

Lesson Plan

Section 3

Objectives *Students will learn that . . .*

1. South Africa ended apartheid and now has a stable government and strong economy.
2. Some countries of Southern Africa have good resources and economies, but several are still struggling.
3. Southern African governments are responding to issues and challenges such as drought, disease, and environmental destruction.

Vocabulary sanctions, Cape Town, enclave

PRETEACH	RESOURCES
___ **What You Will Learn . . .** (SE) Preview Main Ideas, Big Idea, and Key Terms. ___ **Academic Vocabulary** (TE) Review with students the high-use academic term in this section. ___ **If YOU lived there . . .** (SE) Have students read and discuss the section introduction and question.	___ Daily Bellringer Transparency: Section 3 ___ RF: Vocabulary Builder: Section 3 ___ Differentiated Instruction Modified Worksheets and Tests CD-ROM
DIRECT TEACH	**RESOURCES**
___ **Teach the Big Idea Activity** (TE) Students discuss the Main Idea questions and the connection between racism and apartheid. ___ **Main Idea Questions** (TE) Discussion questions ___ **Biography: Nelson Mandela** (SE) ___ **Cross-Discipline Activity: Literature** (TE) Writing an Autobiography ___ **Close-up: Cape Town** (SE) ___ **Critical Thinking: Cause and Effect** (TE) Time Line to Equality ___ **Differentiating Instruction: Special Needs Learners** (TE) South African Countries ___ **Connecting to Economics: Tourism in Southern Africa** (SE) ___ **Critical Thinking: Solving Problems: Ensuring Economic Equality** (TE) ___ **Social Studies Skills: Evaluating a Web Site** (SE)	___ Interactive Reader and Study Guide: Section 3 ___ RF: Geography for Life: Apartheid ___ RF: Critical Thinking: Ending Apartheid: The Economic Impact ___ RF: Biography: Desmond Tutu ___ RF: Primary Source: "Blessed Be the Land of Zimbabwe" ___ RF: Biography: Luisa Diogo
REVIEW & ASSESS	**RESOURCES**
___ **Close** (TE) Briefly review the similarities and differences between the countries of Southern Africa. ___ **Section 3 Assessment** (SE)	___ Quick Facts Transparency: Southern Africa Visual Summary ___ Online Quiz Section 3 (keyword: SGE7 HP8) ___ PASS: Section 3 Quiz

Key: **SE** = Student Edition **TE** = Teacher's Edition **RF** = Resource File **PASS** = Progress Assessment Support System

Lesson Plans for Differentiated Instruction

Africa

Lesson Plans for Differentiated Instruction

The Lesson Plans for Differentiated Instruction provide teachers with strategies to help students meet benchmarks and standards.

The lesson plans provide strategies and suggestions for teaching students with special needs. Differentiated strategies are included for: (1) English-Language Instruction, (2) Special Education Instruction, (3) Advanced Learners/Gifted and Talented Instruction.

Each lesson plan is organized in two sections: (1) a main section, in which strategies appear, and (2) a side column, in which tips and side notes appear.

Main Section—Strategies: For each section in a chapter, at least one strategy is provided for at least two of the three differentiated instruction categories. Each section starts with the section number and title. The next line is the category of instruction being supported. On the next line, the title and recommended time for the strategy is provided. The title indicates how the activity might be used (such as "Prereading") or the focus of the differentiation. Following that is a brief description of the activity. If additional materials are needed for the activity, they are indicated as well. An example of a strategy format follows:

> **SUPPORTING SPECIAL EDUCATION INSTRUCTION**
> **Postreading** (40 minutes)
> **Making a Poster** Have students work in small groups to create a poster about one of the topics in this section. Students' posters should include information and/or illustrations, charts, maps or graphs about Turkish history, the Ottoman Empire, or modern Turkey and its people and culture.

Side Column—Side Notes: For each section in a chapter, one or two side notes that link to a specific strategy or that support differentiated instruction for that section in general are included. These side notes are designed to give teacher tips for using the strategies or for expanding the strategies beyond what is included in the main part of the lesson plan. The side note categories are:

- Culturally Responsive Teaching
- Discipline Connection
- Grammar Tip
- Graphic Organizer
- Language Tip
- Pronunciation Tip
- Teacher Tip
- Technology Tip

Section 1: Geography and Early Egypt

SUPPORTING ENGLISH-LANGUAGE INSTRUCTION

Analyzing Verb Tense (10 minutes)

Understanding the Present The present tense is used in this section in describing some geographical features of Egypt (for example: "The Nile is the longest river in the world"). Draw the following diagram for students:

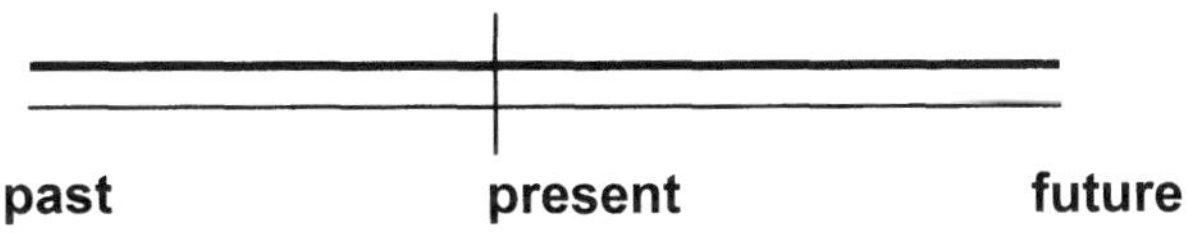

The present is used to describe a fact that was true in the past, true now, and will likely be true in the future. Have students write sentences in the present tense that can describe Egypt in the past, in the present time, and in the future.

SUPPORTING SPECIAL EDUCATION INSTRUCTION

Organizing Information (40 minutes)

Creating a Web Have students create a web graphic organizer like the one at right to help them to process the information from this section. The words "Early Egypt" should be listed in the center circle. Around that, students should draw circles reading "the Nile," "food production," and "the kings." Have students brainstorm words and phrases that relate to these three terms, and add the words around the three circles.

SUPPORTING ADVANCED/GIFTED AND TALENTED INSTRUCTION

Synthesizing Information (45 minutes)

Writing a Children's Book After reading about Menes' rule in early Egypt, have students design a short children's book to help explain his achievements to younger children. Organize the class into small groups and assign a topic to each group. Have each student write one page to summarize the topic. Students should also illustrate the page. Collect all of the pages from the students and bind them into book form. Encourage students to share their book by visiting other classrooms in your school and reading it out loud to the younger students.

Resources

- Spanish Summaries Audio CD Program
- Differentiated Instruction Modified Worksheets and Tests CD-ROM:
  –Vocabulary Flash Cards
  –Vocabulary Builder
  –Section Quiz
  –Chapter Review
  –Chapter Test

Teacher Tip

Provide students with the reading in smaller chunks. You can do this by assigning students to read each sub-section separately.

Graphic Organizer

Concept Web

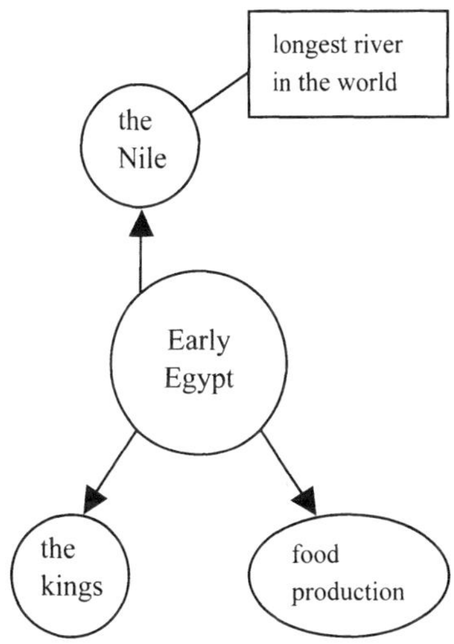

Section 2: The Old Kingdom

SUPPORTING ENGLISH-LANGUAGE INSTRUCTION

Vocabulary Analysis (15 minutes)

Identifying Spatial Vocabulary Read this passage to the class: "At the top of Egyptian society was the pharaoh." Have students write a list of the spatial terms used in describing the development of society in this section (e.g., "at the top, just below," etc.). Using these terms, have students draw a pyramid like the one at right to depict the structure of early Egyptian society.

SUPPORTING SPECIAL EDUCATION INSTRUCTION

Postreading (25 minutes)

Gods of Egypt Chart Have students create a chart of the gods of Egypt listed in the text by folding a piece of paper into three columns. Then have students use a ruler to draw lines between the columns. Students should label the first column "god," the second column "role," and the third column "characteristics." Under each heading, have students list the name of each god (e.g., Amon-Re), the role of the god (e.g., god of the sun), and the characteristics of the god (e.g., king of the gods). As a follow-up assignment, have students investigate the forms of each god on the Internet. A useful Web site can be found at the following address: http://www.ancientegypt.co.uk/gods/explore/main.html.

SUPPORTING ADVANCED/GIFTED AND TALENTED INSTRUCTION

Comparing Information (45 minutes)

Writing a Comparison Essay In this section, students have read about mummies and the afterlife. Have students think about the description of burial practices in early Egypt. Then have them consider the way mummies are portrayed in horror films and television. Have them write a comparison essay in which they discuss the similarities and differences between these two versions of mummies. As a conclusion, have students summarize their thoughts about why mummies are so frightening in American popular culture. Ask students to relate their writing to any movies or TV programs they may have seen that give them ideas about the topic.

Graphic Organizer

Pyramid

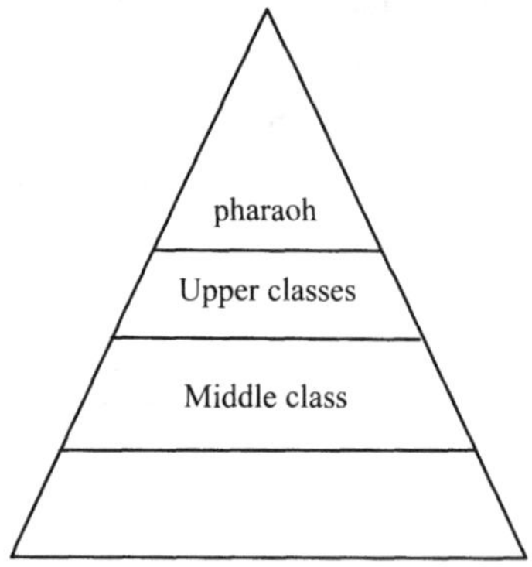

Pronunciation Tip

The word for the Egyptian god "Ptah" is pronounced like this:

TAH

The "P" is silent.

Discipline Connection

Math/Science: Have students investigate the math and science concepts related to the building of the pyramids. They might look for this information in the school library or on the Internet. A good reference site on this topic can be found at the following Web address: http://www.egyptspyramids.com

Section 3: The Middle and New Kingdoms

SUPPORTING ENGLISH-LANGUAGE INSTRUCTION

Vocabulary Analysis (15 minutes)

Identifying Time Words Write a list of time phrases for students on the board (e.g., soon, finally, meanwhile, before, after, first, second, etc.) Have students find where these and other time words are used in this section. Ask them to create a time line of the events described based on the use of these words.

SUPPORTING SPECIAL EDUCATION INSTRUCTION

Paraphrasing (30 minutes)

Family Life in Egypt After reading the section about family life in Egypt, have students brainstorm a list of words and phrases that describe families in early Egypt. Students should then use this information to write a summary paragraph about family life in early Egypt. Encourage students to share their paragraphs with a partner.

Section 4: Egyptian Achievements

SUPPORTING ENGLISH-LANGUAGE INSTRUCTION

Vocabulary Analysis (30 minutes)

Understanding Suffixes Read the following excerpt from the text to the students: "Inside, Egyptian temples were lavishly decorated, as you can see in the drawing of the Temple of Karnak." Explain to students that an adverb can be formed by adding the suffix "-ly" to an adjective:

adjective + -ly = adverb
lavish + -ly = lavishly

List other words with this suffix on the board. Ask students to list any other words they know with this suffix.

SUPPORTING ADVANCED/GIFTED AND TALENTED INSTRUCTION

Interpreting an Object (30 minutes)

Making Inferences Look at the work of artisans, sculptors, builders, carpenters, jewelers, and metal and leather workers pictured in this section. Choose one object and think about the artisan who created it. Take on the role of the artisan and talk about the object with a partner. "What is the purpose or use of your object? To whom are you going to give your object? What were you thinking as you were working on the object? What will be your next project?"

Teacher Tip
Provide students with notes in enlarged type. You can do this on a photocopy machine by enlarging a normal sized sheet of paper.

Teacher Tip
Make a floor map for your classroom. Hang a shower curtain liner on the wall and project an outline map of the Middle and New Kingdoms using an overhead projector. Trace the outline onto the liner with a permanent marker. Label the map with the regions and important sites of battles. Have students "wage war" with the surrounding areas by walking from one area to another on the map.

Discipline Connection
Art: Have students view a variety of artifacts from early Egypt. They can refer to the images in their text, or find more on the Internet or in the school library. One useful Web site can be located at the following address: http://www.mnsu.edu/emuseum/prehistory/egypt/artisans/pottery.html

Ask students to create a replica of their favorite artifact out of papier-mâché or modeling clay.

Section 1: Kush and Egypt

SUPPORTING ENGLISH-LANGUAGE INSTRUCTION

Prereading (10 minutes)

Analyzing Text Structure: Cause and Effect Narration
Brainstorm with students any phrases or words that indicate cause or effect. For example:

- because of this
- as a result
- since
- this led to

List the words and phrases on the board.

Reading (40 minutes)

Analyzing Text Structure: Cause and Effect Narration Have students create a T-chart with the word "cause" written at the top of one column and the word "effect" on the other side. As students read, have them copy down any phrases that indicate effect. Once students read the section, have them go back to the text to fill in the causes for each event in their "effect" column.

Cause	Effect
The Nile's cataracts made parts of the river hard to pass through	Because of this they were natural barriers against invaders

SUPPORTING SPECIAL EDUCATION INSTRUCTION

Postreading (25 minutes)

The Rise and Fall of Kush Have students create a chart detailing the rise and fall of Kush described in the text by folding a piece of paper into three columns. Then have students use a ruler to draw lines between the columns. They should label the first column "The Rise of Kush," the second column "The Height of Kush," and the third column "The Fall of Kush." Under each heading, have students list some important characteristics about each period.

Resources

- Spanish Summaries Audio CD Program
- Differentiated Instruction Modified Worksheets and Tests CD-ROM:
  –Vocabulary Flash Cards
  –Vocabulary Builder
  –Section Quiz
  –Chapter Review
  –Chapter Test

Teacher Tip

Before students read, provide them with a list of questions that they should be able to answer once they have read the section. Check for comprehension after they read by going over the questions with the class.

Technology Tip

Refer students to relevant Web sites that provide additional information about this point in history.

Section 2: Later Kush

SUPPORTING ENGLISH-LANGUAGE INSTRUCTION

Vocabulary Analysis (15 minutes)

Identifying Additive Phrases Brainstorm with students any phrases or words that indicate additional information in the text. For example:

- in addition to
- moreover
- also

List the words on the board. Have students create a flow chart with information from the text.

SUPPORTING SPECIAL EDUCATION INSTRUCTION

Organizing Information (40 minutes)

Creating a Time Line Ask students to brainstorm the major events surrounding the establishment of the early Kush empire. Then ask them to think about the details of the later Kush empire. With this information, have students work in small groups to make a combined time line by organizing the events for early and later Kush in chronological order.

SUPPORTING ADVANCED/GIFTED AND TALENTED INSTRUCTION

Interpreting an Object (30 minutes)

Making Inferences Have students look at the artifacts pictured in this section. Have them choose one object and think about the artisan who created it. Ask them to take on the role of the artisan and talk about the object with a partner. "What is the purpose or use of the object? To whom did they give their object? What was the artisan thinking as they were working on the object? What will be their next project?" Have students write a monologue with the information and share their speeches with the class.

Teacher Tip
Help students to develop mnemonic devices to assist them in remembering the details of an event or time period.

Language Tip
Writing a monologue is different from writing a script for several characters. Have students brainstorm the differences in each type of writing.

Section 1: Empire of Ghana

SUPPORTING ENGLISH-LANGUAGE INSTRUCTION
Vocabulary Analysis (15 minutes)
Understanding Unusual Words Refer students to the discussion of Ghana's beginnings in the text. Explain to students that the phrase "band together" in the sentence "For protection, groups of Soninke families began to band together," means to "form a group or unite" (definition from www.hyperdictionary.com).

SUPPORTING ADVANCED/GIFTED AND TALENTED INSTRUCTION
Interpreting Information (30 minutes)
Creating a Political Cartoon After reading about how the rulers in Ghana forced traders to pay excessive taxes, have students design a political cartoon that illustrates the topic. They should include some form of illustration (hint: encourage the use of computer graphics if students prefer not to draw) along with a speech bubble and/or caption. Remind students that their cartoon doesn't necessarily have to be funny. It should, however, depict this time in history in some novel way. Have students share their cartoons with the class.

Section 2: Mali and Songhai

SUPPORTING ENGLISH-LANGUAGE INSTRUCTION
Vocabulary Analysis (15 minutes)
Identifying Changing Phrases Brainstorm with students any phrases or words that indicate a change of information in the text. For example:

- nonetheless
- despite
- however

List the words on the board. Have students create a comparison chart with information from the text.

SUPPORTING SPECIAL EDUCATION INSTRUCTION
Reviewing Information (40 minutes)
Using Self-Questioning Techniques Have students use self-questioning techniques to review the material addressed in this section. As students read, encourage them to write questions about the material they do not understand from the reading on a separate piece of paper. Students should then meet in small groups to share their questions and look for answers in their textbooks.

Resources

- Spanish Summaries Audio CD Program
- Differentiated Instruction Modified Worksheets and Tests CD-ROM:
  –Vocabulary Flash Cards
  –Vocabulary Builder
  –Section Quiz
  –Chapter Review
  –Chapter Test

Teacher Tip

Begin a lesson about the topic by "previewing" some of the key vocabulary from the section. Find small objects or photos representing the vocabulary. Place these items in a bag or box with a lid. Have students pull the items from the bag one by one; discuss the object and provide a short definition/explanation for each one.

Teacher Tip

Provide students with notes that are color-coordinated, using different colors to distinguish between ideas.

Culturally Responsive Teaching

Have students think about the topic of slavery discussed in the section. Ask them to think about what they know about slavery in the modern world.

SUPPORTING ADVANCED/GIFTED AND TALENTED INSTRUCTION

Synthesizing Information (30 minutes)

Writing a Screenplay The life of Sundiata has all the elements of a Hollywood movie – action, battles, and even magic! Have students brainstorm what a movie about Sundiata's life might be like. Have them design a storyboard with several scenes that would appear in the film. Students should include the characters, the setting, and sample dialogue for each scene.

Section 3: Historical and Artistic Traditions

SUPPORTING SPECIAL EDUCATION INSTRUCTION

Reviewing Information (20 minutes)

Identifying Main Ideas Help students look for key words in the text. For example:

- the following
- the most important point

Have students underline or highlight the main points in this section. Then have students summarize each main idea in a sentence or two, or by drawing a picture of the main idea.

SUPPORTING ADVANCED/GIFTED AND TALENTED INSTRUCTION

Extending Information (30 minutes)

Telling a Story Have students take on the role of a *griot.* Ask them to think of a story they may have read recently. Have them memorize the story and think about ways of incorporating facial expressions, gestures, and props. Have students tell their story to the entire class in a storytelling session.

SUPPORTING ADVANCED/GIFTED AND TALENTED INSTRUCTION

Synthesizing Ideas (30 minutes)

Writing an Epic Poem After reading about epic poems, have students brainstorm a story about a hero that is important to them. Encourage students to think about the struggles that their hero may have gone through in her/his life. Have students write a shortened version of an epic poem in which they include historical events and other pertinent information mixed with songs of praise. Ask students to read the poem in front of the class.

Technology Tip

Use the overhead projector to present outlines and notes to the class. As you lecture or discuss a topic, point to the corresponding topic on the overhead transparency.

Teacher Tip

Allow students to share notes from your lecture or readings. They may photocopy a peer's notes.

Culturally Responsive Teaching

After reading about proverbs in this section of the text, have students research proverbs that come from their own cultural background. Have students share their proverbs with the class. They can also write out their proverbs, decorate them, and hang them around the classroom with the country/culture of origin labeled.

Technology Tip

Design a "Webquest," or Internet-based scavenger hunt, for students around the information in this section. Choose four or five "treasures" that students must find, and then have students search for the items on the Web.

Section 1: Physical Geography

SUPPORTING ENGLISH-LANGUAGE INSTRUCTION

Vocabulary Analysis (10 minutes)

Understanding Word Groups List a word group that students may be familiar with on the board. For example:

study, student, studious

Have students copy the words and underline the similar part. Ask students to explain the meaning of each word. If students do not know some of the words in the list, have them guess the meaning based on the words they do know. Confirm correct meanings.

Reading (15 minutes)

Understanding Word Groups In the text, students will encounter the words *fertile* and *fertilizer*. Have students use context clues, hints in the sentence or sentences surrounding a word, to determine their meanings. For example, from "Silt is finely ground fertile soil good for growing crops," they might determine that fertile is good for growing things. Have students try to determine the meaning of the word *fertilizer*, found later on the page, based on their understanding of the word *fertile*. Have them use both words in sentences.

SUPPORTING SPECIAL EDUCATION INSTRUCTION

Gathering Information (30 minutes)

Creating a Concept Map Draw a concept map like the one shown at right on the board. Read Section 1 aloud with students, pausing frequently to ask them to gather information in the chart. You may wish to begin by reviewing ideas for gathering and organizing information, such as using subheads as labels. In this case, each outer circle could be labeled with a subhead from the "Physical Features" section. Have students gather important details from each section.

Summarizing Information (30 minutes)

Writing a Summary Remind students that a summary includes the main ideas, but not all the detail of the original. Have students write a summary entitled "The Physical Geography of North Africa." Students may prefer to draw a picture to summarize what they have learned. Encourage them to label their pictures to reinforce place names and key vocabulary terms.

Resources

- Spanish Summaries Audio CD Program
- Differentiated Instruction Modified Worksheets and Tests CD-ROM:
  –Vocabulary Flash Cards
  –Vocabulary Builder
  –Section Quiz
  –Chapter Review
  –Chapter Test

Language Tip

Extend the word lesson on *fertile* and *fertilizer* by working with antonyms. Explain that antonyms are opposites. Give common examples, such as *easy* and *hard* or *wet* and *dry*. Point out the word *barren* in the section "The Sinai and the Suez Canal," and tell students that it is an antonym for *fertile*. Have students guess the meaning of *barren* based on their understanding of *fertile*. Then, have them write a compare and contrast sentence using the antonyms.

Graphic Organizer

Concept Map

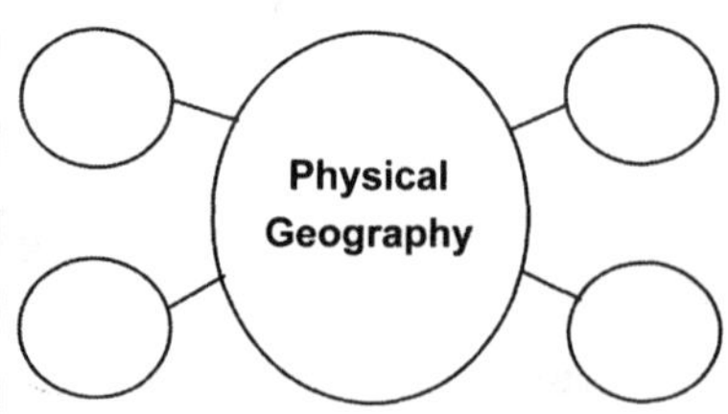

Section 2: History and Culture

SUPPORTING ENGLISH-LANGUAGE INSTRUCTION

Analyzing Verb Tenses (10 minutes)

Understanding the Past Tense Read the following excerpt to students: "The early Egyptians used water from the Nile to grow wheat, barley, and other crops." Draw the following diagram for students.

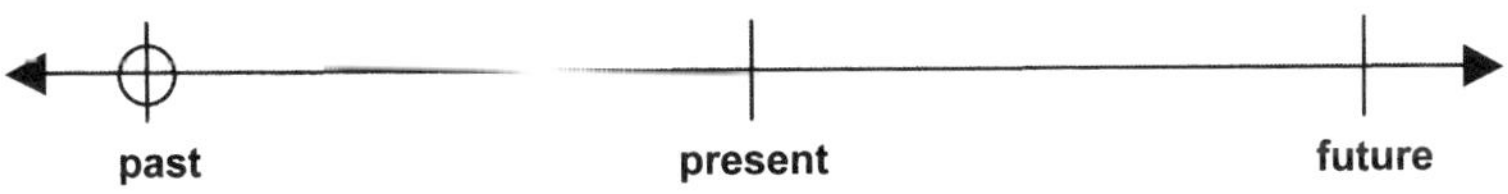

Explain to students that the sentence is stated in the past tense. Have students think of other examples using the same structure.

SUPPORTING SPECIAL EDUCATION INSTRUCTION

Reading (30 minutes)

Identifying Difficult Passages As students read this section for the first time, have them place sticky notes on the pages next to passages that are confusing or that they do not understand. On each sticky note, have students write a question or a vocabulary word that they are having difficulty with. Once they finish reading, have them work in small groups to try to clear up any questions. Check in with groups and help them to answer any unresolved questions by directing them to dictionaries or by conferencing with them. Have students reread the passage to see if their questions have been answered.

SUPPORTING ADVANCED/GIFTED AND TALENTED INSTRUCTION

Expressing Information (40 minutes)

Writing a Film Script If you were a filmmaker working on a documentary about North Africa or set in North Africa, what images would you use to capture the setting? Write a description of what the viewer would see in the opening shots. Draw upon the images and information in your textbook for ideas, and use the Internet to find other appropriate images. Write a brief summary of the main ideas and purpose of your film and a description of the opening shot. Include details about what cities, monuments, or other landmarks you would show. Also describe the sights and sounds in detail.

Grammar Tip

Students will encounter some present tense sentences in this history section, such as "The average weight of each of the 2.3 million blocks in the Great Pyramid is 2.5 tons." Explain that sentences in the present tense indicate that something exists or is happening now. In this case, the blocks still weigh 2.5 tons.

Culturally Responsive Teaching

To help students see that the text shows only a sampling of food and customs, ask them to write a brief summary of the food and customs of the United States. Have students share their summaries with the class and notice differences in what they found representative or important.

Technology Tip

Students may find images to support their scripts at sites such as http://www.africa-ata.org or by searching for "North Africa" or any of the countries in this region at http://www.nationalgeographic.com

Section 3: North Africa Today

SUPPORTING ENGLISH-LANGUAGE INSTRUCTION

Reading (30 minutes)

Organizing Information Have students create a three-column chart about North Africa by folding a sheet of paper into three columns. Then have students use a ruler to draw lines between the columns. Have students label the first column "Country," the second column "Government and People," and the third column "Resources, Industry, and Challenges." Have students list the countries of North Africa in the first column, with the names spread out along the length of the page. As students read, have them enter details about the government and population of the each country in the second column and about resources, industry, and challenges in the third column.

Postreading (20 minutes)

Organizing Information To reinforce ideas from the chapter, ask students to find similarities and differences between countries. Have them refer to the charts they made to answer questions you ask. For example, ask, "Which countries have similar resources?" or "What countries struggle with the role Islam will play in their government?" Questions could also take this format: "Name something Egypt and Algeria have in common" or "What is unique geographically about Morocco?"

SUPPORTING ADVANCED/GIFTED AND TALENTED INSTRUCTION

Interpreting Information (30 minutes)

Creating a Political Cartoon After reading about the challenges of North Africa today, have students design a political cartoon that illustrates one challenge of the region or a country in the region. They should include some form of illustration and speech bubbles or a caption. (Students who prefer not to draw may use computer graphics.) Remind students that their cartoon doesn't have to be funny. It should, however, depict an issue in contemporary North Africa in a novel way. Have students share their cartoons with the class.

Teacher Tip

Use the section to review cause and effect. On the board or chart paper, make a simple T-chart with the word "cause" at the top of the left column and the word "effect" at the top of the right. Give an example of cause and effect and write it in the chart. For example: Cause—People move to Cairo but there is not enough housing. Effect—People live in slums, on boats, or in cemeteries. Have students find other causes and effects as they read.

Discipline Connection

Language Arts: Extend the political cartoon activity and give students another opportunity to practice writing by having them write a letter to the editor about one of their classmate's political cartoons. First, as a class, read and discuss some letters to the editor. Then have students write their own letters. Students should respond to the cartoon and express their own views (agreeing or disagreeing with the cartoon) on the issue.

Language Tip

Political cartoons and political articles both express ideas about issues. Have students brainstorm the differences between writing an article on a political topic and creating a political cartoon.

Section 1: Physical Geography

SUPPORTING ENGLISH-LANGUAGE INSTRUCTION

Vocabulary Analysis (15 minutes)

Identifying Changing Phrases On the board, write the following sentence from the text: "Although the Sahel is quite dry, it does have enough vegetation to support hardy grazing animals." Explain to students that *although* is an example of a changing phrase. It indicates a change of information in the text. Give other examples of changing phrases, such as *however* and *despite*. Help students identify other sentences in the text that include changing phrases and brainstorm some of their own. Then have students break down the information in these sentences using a comparison chart like the one at right.

SUPPORTING SPECIAL EDUCATION INSTRUCTION

Create a Chapter Outline (40 minutes)

West Africa—A Virtual Tour Have students imagine a computer program that would allow users to "fly" over the landscapes of West Africa. Ask students to review the text and create a menu for this program, organized into two sections: "Physical Features" and "Climate Regions." Under each section heading, users would be able to click on various options. For example, "coastal plains" might be one option under Physical Features. "Desert" and "steppe" are examples of options that might appear under Climate Regions. Encourage students to compare their menus in small groups.

SUPPORTING ADVANCED/GIFTED AND TALENTED INSTRUCTION

Writing Poetry (40 minutes)

Ode to West Africa Have students choose a geographical region or feature described in the text and imagine they live there. How would they express their pride in their homeland through poetry? Have them write an ode to the area (inland plains, the Niger, the Sahel, etc.), addressing the region as "you" to make it personal.

Resources

- Spanish Summaries Audio CD Program
- Differentiated Instruction Modified Worksheets and Tests CD-ROM:
  –Vocabulary Flash Cards
  –Vocabulary Builder
  –Section Quiz
  –Chapter Review
  –Chapter Test

Graphic Organizer

Comparison Chart

Fact 1	*Fact 2*
The Sahel is quite dry.	It has enough vegetation to support hardy grazing animals.

Teacher Tip

You may choose to introduce the poetry assignment by presenting students with some existing odes. Find examples on www.poets.org

Section 2: History and Culture

SUPPORTING ENGLISH-LANGUAGE INSTRUCTION

Prereading (10 minutes)

Analyzing Text Structure: Sequential Narration Explain to students that much of this section of the chapter is narrated in a sequential fashion. It explains events in West African history in sequence, or the order in which they occurred. One way sequential order can be identified is by the use of dates or approximate dates (such as "about 1300"). Have students look through the text and write a list of approximate dates that appear in the section. They should list these dates down the left-hand side of a sheet of paper.

Reading (40 minutes)

Analyzing Text Structure: Sequential Narration As students read, have them write a brief sentence or two that describe what happened on each of the dates that they listed in the previous activity. When students have completed their lists, have them create a time line with the information.

SUPPORTING ADVANCED/GIFTED AND TALENTED INSTRUCTION

Interpreting Information (30 minutes)

Creating a Political Cartoon Have students design a political cartoon illustrating European access to West Africa's resources. They should include some form of illustration (hint: encourage the use of computer graphics if students prefer not to draw) along with a speech bubble and/or caption. Remind students that their cartoon should imply an opinion or personal view on the European colonization of West Africa. Have students share their cartoons with the class.

Pronunciation Tip

The word "Timbuktu" is pronounced like this:

Tim•buck•TOO

The first "u" is short, while the second is long.

Discipline Connection

Mathematics: Have students use their time lines to determine approximate time spans of, and between, certain events. Where possible, have them calculate time spans in terms of years, decades, and centuries.

Section 3: West Africa Today

SUPPORTING ENGLISH-LANGUAGE INSTRUCTION

Analyzing Verb Tense (15 minutes)

Understanding the Present Tense Read the following sentences from the text to students:

"Senegal wraps around Gambia."
"Now, both Liberia and Sierra Leone are trying to rebuild."

Explain to students that these sentences demonstrate forms of the present tense. They both describe something that is happening, or is true, right now. Have students identify other sentences in the text that follow this general structure.

SUPPORTING SPECIAL EDUCATION INSTRUCTION

Organizing Information (30 minutes)

West African Countries Chart Have students create a chart of either West Africa's coastal countries or its Sahel countries. They can start by folding a piece of paper into three columns. Then have them use a ruler to draw lines between the columns. Students should label the first column "country," the second column "economy," and the third column "history/other." Have students list the name of each country (e.g., Gambia), the country's main economic resource or resources (e.g., peanuts, tourism), and at least two historical or other details about the country (e.g., surrounded by Senegal, Wolof language spoken) in the appropriate columns.

Organizing Information (20 minutes)

Writing a Summary Paragraph Using information from the previous activity, have students write a paragraph briefly describing the countries in their chosen region. Remind students that the paragraph should have a main idea, which may be stated as an opening or closing sentence. Supporting details that give more information about the main idea should fill out the rest of the paragraph.

Grammar Tip

A sentence in which the verb is one word, with "s," "es," or no letters added (first example to the left), is in the *simple present tense.* When the verb begins with the word "is" or "are" and is followed by a verb ending in "-ing" (second example), the sentence is in the *present continuous tense.*

Technology Tip

If students need more information on a particular country in order to complete their chart, have them find additional facts online. The Web site of an African studies center, such as the University of Pennsylvania's (http://www. africa.upenn.edu/Home_Page/Country.html), may be a good place to start.

Section 1: Physical Geography

SUPPORTING ENGLISH-LANGUAGE INSTRUCTION

Vocabulary Analysis (20 minutes)

Understanding Word Groups The text uses the word "location" when describing Mount Kilimanjaro's proximity to the equator. Explain to students that *location* comes from the Latin word *locus,* meaning *place.* List other words, such as *locate* and *local,* that are formed from the same root. Give students examples of other words that belong to word groups, and ask if they can identify more words in those groups.

SUPPORTING SPECIAL EDUCATION INSTRUCTION

Organizing Information (30 minutes)

A Brainstorming Web Have students brainstorm the main ideas and topics in this section of the reading. Ask students to think of all the terms and ideas from the section and write these words on the board in a web, using the sample web at right as a model.

Organizing Information (30 minutes)

Writing a Letter Ask students to imagine they are archaeologists working in East Africa. They have been there several months and have seen the entire region. Have students write a letter to their family using the information from the previous activity. Encourage them to share as much about the topic as possible in the letter. Some students may prefer to draw a picture to summarize what they have learned.

SUPPORTING ADVANCED/GIFTED AND TALENTED INSTRUCTION

Interpreting Information (20 minutes)

Writing a Compare/Contrast Essay The text describes various regions of East Africa in terms of their physical features and climates. Examples include mountains, highlands, plains, and tropical savannas. Have students write an essay about two of these regions in which they discuss the similarities and differences between them. As a conclusion, have students summarize their thoughts about East Africa's spectacular landscape.

Resources

- Spanish Summaries Audio CD Program
- Differentiated Instruction Modified Worksheets and Tests CD-ROM:
  –Vocabulary Flash Cards
  –Vocabulary Builder
  –Section Quiz
  –Chapter Review
  –Chapter Test

Graphic Organizer

Brainstorming Web

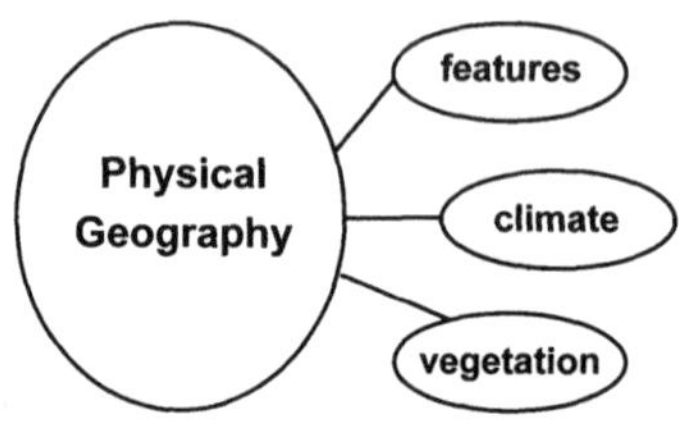

Language Tip

Given the nature of the subject matter, this compare/contrast essay could also be considered a descriptive one. Encourage students to use several "colorful" adjectives and phrases when describing their chosen regions.

Section 2: History and Culture

SUPPORTING ENGLISH-LANGUAGE INSTRUCTION

Prereading (10 minutes)

Implied Cause and Effect Read the following sentence to the students: "As East Africans traded with Arabic speakers from Southwest Asia, Swahili developed." Explain to students that cause-and-effect relationships are not always described with phrases like "because of" and "as a result of." Often, causality is only implied. Have students brainstorm other examples that imply a cause-and-effect relationship.

Reading (10 minutes)

Cause and Effect Have students create a T-chart with the word "cause" written at the top of one column and the word "effect" on the other side. As students read, have them copy down any phrases that indicate effect, or something having happened. Once students have read the section, have them go back to the text to fill in their "cause" column with causes for each effect.

Cause	Effect
East Africans traded with Arabic speakers from Southwest Asia.	Swahili developed.

SUPPORTING ADVANCED/GIFTED AND TALENTED INSTRUCTION

Interpreting Information (40 minutes)

Writing a Skit Divide students into small groups. Have each group choose one event described in the section and write a short skit depicting the event. Each group member should play a role in the skit, although not everyone has to have speaking lines. Encourage students to incorporate as many factual details as possible within their skit, but allow fictional "filler" as needed. Have each group perform their skit for the rest of the class. Encourage students to make props for their skits.

Grammar Tip

Once students have completed the cause-and-effect exercises, have them rewrite each example so that causes and effects are clearly stated. For example, the sentence to the left might become: "East Africans traded with Arabic speakers from Southwest Asia. As a result, Swahili developed."

Technology Tip

Bring a video camera into the classroom. Place each group in charge of videotaping another group's performance. Encourage teamwork by suggesting one student handle the camera while another student quietly gives directions (zoom in, get a certain angle, etc.).

Section 3: East Africa Today

SUPPORTING ENGLISH-LANGUAGE INSTRUCTION

Vocabulary Analysis (20 minutes)

Identifying Additive Phrases Read the following excerpt from the text to students: "Somalia has been troubled by violence in the past. In addition, the country has often had no central government of any kind." Explain to students that the phrase *in addition* is an additive phrase. It indicates additional information in the text. Other additive phrases include:

- also
- moreover
- further

List these phrases on the board. As students read the section, have them look for sentences containing additive phrases. Have them rewrite these sentences using alternative additive phrases.

SUPPORTING SPECIAL EDUCATION INSTRUCTION

Making Lists (15 minutes)

Economic Resources After students have read the section, have them make a bulleted list of East Africa's major sources of income. The list may include industries and natural resources. After they have created the list, have students identify which resources are significant in only one East African country. Have them write the country's name in parentheses next to the resource.

SUPPORTING ADVANCED/GIFTED AND TALENTED INSTRUCTION

Synthesize Ideas (40 minutes)

Writing a Real Estate Ad Tell students that they are real estate agents in an East African country of their choice. Have them write an ad for a home they want to sell. They should focus on the location of the home, offering details that would make someone want to move there. For example, a home right within a bustling city would be perfect for business commuters. The ad should also try to "sell" the country itself, for potential foreign buyers. Encourage students to read their ad aloud to a classmate.

Grammar Tip

As students use alternative additive phrases to rewrite sentences, remind them to make sure that subjects and verbs agree. For example: The **coach**, not to mention the players, **wants** to start the game.

Discipline Connection

Science: Have students choose one of the crops or natural resources in their list and learn more about it. What conditions make a particular crop easy to grow? How is a certain mineral mined? Have students write a paragraph or brief essay on the topic.

Teacher Tip

Collect students' ads to create a real estate section of a fictional East African newspaper. Post the section in the classroom.

Section 1: Physical Geography

SUPPORTING ENGLISH-LANGUAGE INSTRUCTION

Vocabulary Analysis (10 minutes)

Understanding Prefixes This section of text contains several words with prefixes. Examples include:

- impossible
- transportation
- promote

Write these words on the board. Ask students if they can define each word and identify the prefix in each. Help students determine the meaning of the prefix based on the word's definition.

SUPPORTING SPECIAL EDUCATION INSTRUCTION

Reinforcing Vocabulary (30 minutes)

Understanding Geographical Terms Have students create a personal dictionary using vocabulary words from the section. Some words they might define include:

- plateau
- rapids
- interior
- latitude
- tropical

Encourage students to look for multiple definitions of these terms, including information from the text and from the glossary. Have students write sentences demonstrating that they understand each term.

SUPPORTING ADVANCED/GIFTED AND TALENTED INSTRUCTION

Synthesizing Information (20 minutes)

Writing a Children's Book The text describes several animals found in the forests of Central Africa. Have students write a children's storybook featuring the animals of this region. Encourage them to draw pictures or find images online to go with their story. Have them "bind" their book with staples, string, or fasteners.

Resources

- Spanish Summaries Audio CD Program
- Differentiated Instruction Modified Worksheets and Tests CD-ROM:
  –Vocabulary Flash Cards
  –Vocabulary Builder
  –Section Quiz
  –Chapter Review
  –Chapter Test

Pronunciation Tip

The word "plateau" is pronounced like this:

pla•TOH

The letters "eau" are generally pronounced with an "oh" sound. A word with these letters originates from the French language.

Language Tip

Before students write, explain some basic literary elements. For example, their story should have characters, a plot, and possibly a conflict. Explain that if their animals talk, they are using the element of personification.

Section 2: History and Culture

SUPPORTING ENGLISH-LANGUAGE INSTRUCTION

Reading (20 minutes)

Taking Notes As students read the first part of the section ("History"), have them take notes. Suggest that they jot down main events described, including key dates, names, and places. Explain that generally notes are for personal use. This means complete sentences are not necessary, abbreviations are encouraged, and style of organization (outline, bulleted list, etc.) is a matter of preference.

Postreading (15 minutes)

Creating a Flow Chart Using information from the previous activity, have students make a flow chart like the one on the right. The chart should show the general order of notable events in Central African history. Have students place the events in separate boxes and use the arrows to show how one event led to, or came after, another.

SUPPORTING SPECIAL EDUCATION INSTRUCTION

Postreading (15 minutes)

Paraphrasing Once students have read the section, have them decide which piece of Central Africa's history and culture they find the most interesting. Have them rewrite that part of the text in their own words.

SUPPORTING ADVANCED/GIFTED AND TALENTED INSTRUCTION

Making Inferences (30 minutes)

Writing a Radio Ad Ask students to imagine that a group of musicians from Central Africa will be performing at your school. Have students write a radio ad to promote the event throughout your community. As they review relevant information in the text, have them think about what the various musical instruments might sound like. Encourage them to incorporate descriptive details that would make people want to attend the performance.

Culturally Responsive Teaching

Review with students the non-English languages spoken throughout Central Africa. Ask students if they speak, read, or understand any of these languages, or if they know someone who does.

Graphic Organizer

Flow Chart

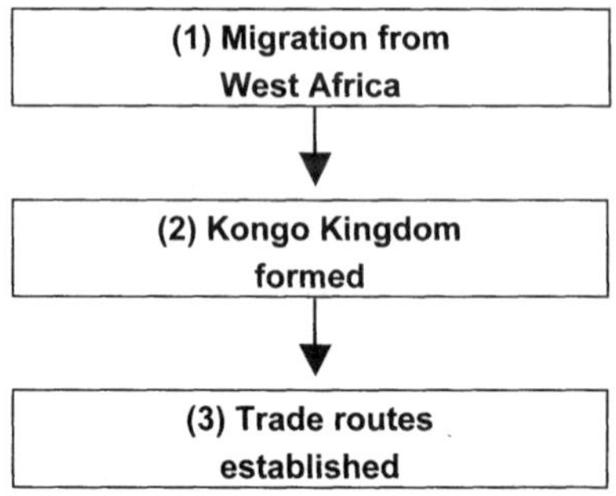

Discipline Connection

Art and Music: Obtain one or more of the following items to present to the class: a Central African sculpture, mask, or garment (or a photograph of one); a musical instrument popular in Central Africa; an audio or video recording of traditional Central African music.

Section 3: Central Africa Today

SUPPORTING ENGLISH-LANGUAGE INSTRUCTION

Analyzing Verb Tense (15 minutes)

Understanding the Future Tense Read the following sentence from the text to students: "People who sleep under these nets will be protected from mosquitoes and malaria." Draw the following diagram for students:

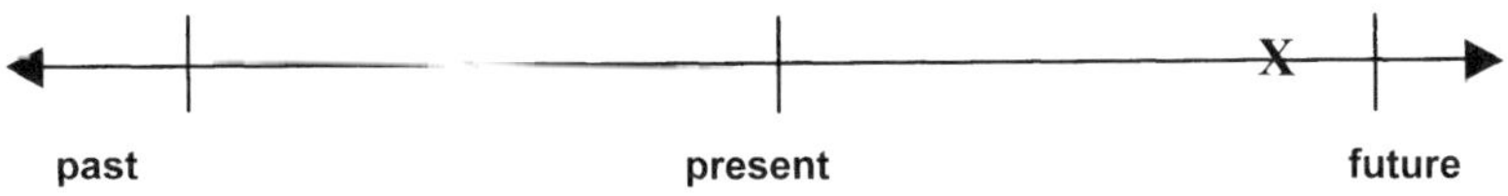

Explain to students that the sentence is stated in the future tense. It describes something that is expected to happen. Have students think of other examples using the same structure.

SUPPORTING SPECIAL EDUCATION INSTRUCTION

Prereading (10 minutes)

Predicting Main Ideas Before students read, have them look at the headings, subheadings, and highlighted words in this section. They should also pay attention to the photos and other graphics that appear within the reading. Based on this information, ask them to predict what they will learn about countries in Central Africa when they read the section. Have them share their predictions with a partner. As they read, encourage students to determine whether their predictions were correct.

Postreading (30 minutes)

Working in a Study Group Have students meet in groups of three or four people to discuss their experiences with the previous activity. Encourage them to listen while each student shares his or her predictions about the section and what turned out to be true. Have students discuss why they made the predictions they did.

Teacher Tip

Provide students with a copy of your lecture or presentation notes for their review and for use as a study guide.

Technology Tip

Offer students additional practice predicting main ideas by giving them the URL for an educational Web site, such as http://www.kidshealth.org, your local newspaper's Web site, or another appropriate site. Based on the home page, what type of organization do students think the site represents? What kinds of information will they find on the site? Repeat the activity with two or three other sites.

Section 1: Physical Geography

SUPPORTING ENGLISH-LANGUAGE INSTRUCTION

Prereading (10 minutes)

Analyzing Text Structure: Descriptive Narration Explain to students that this section of the chapter is narrated in a descriptive fashion. The section describes the physical geography of Southern Africa. Have students identify chunks of text that indicate descriptive narration. Examples from the section include:

- "the land rises sharply from a narrow coastal plain"
- "it forms a swampy inland delta"
- "the eastern slopes of these mountains are rainy"

Write examples on the board.

Reading (40 minutes)

Analyzing Text Structure: Descriptive Narration Have students create a T-chart with the words "Place or Feature" written at the top of one column and the word "Description" written on the other side. As students read, have them write down places or geographical features mentioned in the text. Once they have read the section, have them fill in the "description" column with words and phrases that describe each place or feature.

Place or Feature	Description
Augrabies Falls	rocky, several miles wide

SUPPORTING SPECIAL EDUCATION INSTRUCTION

Describing a Place (30 minutes)

Postcards from Southern Africa Tell students they are on vacation somewhere in Southern Africa. Have them create a postcard to send to someone back home. The front of the postcard should have an image representing the region in which they are staying or one they have visited. Students may draw the image or use a photo. They may find appropriate images on travel Web sites promoting African tourism. On the back of the card, students should write a brief note explaining what the image shows. Encourage them to include as many details as will fit on the left side.

Resources

- Spanish Summaries Audio CD Program
- Differentiated Instruction Modified Worksheets and Tests CD-ROM:
  –Vocabulary Flash Cards
  –Vocabulary Builder
  –Section Quiz
  –Chapter Review
  –Chapter Test

Language Tip

Explain to students that descriptive narration often compares and contrasts subjects. Words such as "in contrast to" and "unlike" indicate this technique. In some cases, comparison is less direct, and no key words are used. Help students identify examples of comparison/contrast in this section of text.

Teacher Tip

To begin the postcard activity, provide students with blank, pre-cut pieces of card stock or index cards (approximately 4"x 6"). Have them draw a line (using a ruler) down the center of one side, to create spaces for text and an address.

Section 2: History and Culture

SUPPORTING ENGLISH-LANGUAGE INSTRUCTION

Vocabulary Analysis (15 minutes)

Identifying Proper Nouns Write the following terms from the text on the board:

- Shona
- Great Zimbabwe
- Namib Desert
- Asia
- Christianity

Explain to students that these are proper nouns. They are names of specific people, places, or things, and should always be capitalized. Use these and other examples from the text to describe various types of proper nouns (cultural groups, cities, religions, etc.). Help students think of other types of proper nouns not represented in the section, along with examples of each.

SUPPORTING SPECIAL EDUCATION INSTRUCTION

Reading (30 minutes)

Self-Questioning Techniques Have students use self-questioning techniques to review the material addressed in this section. Encourage students to write down questions that come to their mind as they read. When they have finished reading, have students meet in small groups to share their questions and look for answers in their textbooks.

SUPPORTING ADVANCED/GIFTED AND TALENTED INSTRUCTION

Interpreting Information (40 minutes)

Writing an Interview Ask students to imagine it is 1985. Apartheid laws are still in effect in South Africa. Have students write an interview with a member of one of South Africa's non-white groups (blacks, Coloureds, or Asians). Encourage students to ask questions that reveal how apartheid affects the person's life. How do his or her rights and freedoms differ from white people's? What changes would the person like to see?

Grammar Tip

Adjectives formed from proper nouns are called proper adjectives. For example, a native of *Asia* may be called *Asian*. Someone who practices *Christianity* is *Christian*. "Christian" is an example of a term that can be used as either a proper adjective or a proper noun ("a Christian").

Culturally Responsive Teaching

Have students write an essay or journal entry expressing their thoughts on apartheid. Encourage them to frame their writing in terms of their own ethnicity and cultural descent. Let students know they do not have to submit their writing if they prefer to keep it private.

Section 3: Southern Africa Today

SUPPORTING ENGLISH-LANGUAGE INSTRUCTION

Analyzing Verb Tense (15 minutes)

Understanding the Past Perfect Tense Read the following sentence from the text to students: "This group had been formed to protect the rights of black South Africans." Explain to students that the sentence is stated in the *past perfect* tense. It describes something that happened in the past, but is relevant to a current event or discussion. Have students think of other examples using the same structure.

SUPPORTING SPECIAL EDUCATION INSTRUCTION

Organizing Information (30 minutes)

A Brainstorming Web Have students brainstorm the main ideas and topics covered in this section. Ask students to express these using one-word terms or very short phrases. Write the terms and phrases on the board in a web, using the sample web at right as a model.

Organizing Information (30 minutes)

Writing a Summary Paragraph Using the information from the previous activity, have students write a short paragraph entitled "Southern Africa Today." Encourage students to include as much information about the topic as possible. Some students may prefer to draw a picture to summarize what they have learned.

SUPPORTING ADVANCED/GIFTED AND TALENTED INSTRUCTION

Synthesizing Information (40 minutes)

Writing a Feature Story The text calls Botswana "one of Africa's success stories." Have students learn more about how and why the country has earned this description. Then, have them write a feature story on Botswana for a travel or geography magazine. Encourage students to find information through printed sources and the Internet. Students can visit the official Web site of the Republic of Botswana at www.gov.bw.

Grammar Tip

The past perfect tense is formed when the verb in a sentence contains *had* plus a past participle verb form (*lived, gone, broken,* etc.). In the past perfect tense, adverbs sometimes break up the verb:

He had never flown before.

I had always loved that song.

Graphic Organizer

Brainstorming Web

Interactive Reader and Study Guide Teaching Guide and Answer Key

Africa

Interactive Reader and Study Guide

The *Interactive Reader and Study Guide* provides a chapter reading strategy and summaries of the main ideas of each section. The goal of this *Intereactive Reader and Study Guide* is to provide a written summary of each section in the textbook for students who are having problems accessing the concepts that they must master for assessment. The guide should help students understand the big ideas or main concepts of each section.

The *Interactive Reader and Study Guide* provides students with opportunities to interact with the text, which allows them to access the power and meaning of the printed word. Interaction with the text should help students develop positive attitudes about reading.

The structure of the activities allows students to warm up by looking at a chapter summary graphic organizer that summarizes the main ideas in the chapter. These organizers are based on the Big Ideas and Main Ideas listed at the start of each section.

Each graphic organizer is followed by two to four Comprehension and Critical Thinking questions. These questions are related to the material in the graphic organizer. Questions frequently direct students to complete part of the graphic organizer or to add to it. Sometimes, students are instructed to complete the graphic organizer and then answer the questions.

Each of the section summaries has features to help students access the main ideas of the section. These features include:

- **Main Ideas Box** This is the text of the section main ideas from the section opener page in the Student Edition chapter.
- **Key Terms and Places** This is the list of key terms and places from the Student Edition section opener page. Each term has a short definition that matches the definitions provided in the running text of the Student Edition. The definitions have been shortened if necessary.
- **Section Summary** This is a summary of the section text. The summary covers the section's main ideas, addresses the power standard, and includes all boldfaced key terms and places in the section. The section heads from the student text are used to organize the summary.
- **Call-Out Box Questions and Activities** These activities appear in the side column next to the summary. The questions and activities are written to encourage students to interact with the text. These questions and activities may provide additional information related to a portion of the summary, ask students to circle or underline specific parts of the summary, or ask questions about the summary. The questions and activities vary both in type and difficulty level. They help reinforce, explain, clarify, or extend the main ideas in the section. These questions and activities are designed to be keys to help students unlock the text—and thereby help students who have difficulty accessing the concepts to master the information they need for assessment.
- **Challenge Activity** This activity follows the section summary. It is a short activity that challenges students and builds off of one or more of the section's main ideas. This Challenge Activity is usually aimed at advanced and gifted and talented students.

When students have finished the Chapter Opener and Section Summary pages for a chapter, they should have a grasp of the main ideas in the section—the concepts that will be assessed on standardized tests. The *Interactive Reader and Study Guide* can be used by a teacher in a variety of ways, including

- as a way to review and reinforce the regular student textbook,
- as a way to give struggling students more time to focus on only the most important parts of the section, which will give them the chance to master the material and be successful when assessed, and
- as a way to give advanced or gifted and talented students an opportunity to master the basic material quickly so that they might move on to an extension activity.

History of Ancient Egypt

COMPREHENSION AND CRITICAL THINKING

1. around 3100 BC
2. It marked the start of the period known as the Old Kingdom.
3. Egypt was stable for about 300 years.
4. Egyptian leaders took control of all possible invasion routes into Egypt.

SECTION 1

Call-Out Boxes

1. It brings water and fertile soil to a desert land.
2. to make Egypt stronger and more stable

Challenge Activity

Students' speeches should include these points of persuasion: the ability to guarantee a steady food supply, markets for goods, peace.

SECTION 2

Call-Out Boxes

1. No, because the leader was seen as both a political and religious figure.
2. the lower class
3. a person's life force

Challenge Activity

Students' essays may mention that Egyptians viewed the pharaoh as an all-powerful ruler and a divine being, whereas a U.S. president is the elected leader of a democracy.

SECTION 3

Call-Out Boxes

1. Southwest Asia
2. south
3. architects
4. No, farmers continued to work hard and pay taxes and duties.

Challenge Activity

Students' "want ads" may include the following professions: scribe, artisan, artist, architect, trader, farmer. Students' letters should include elements of persuasive writing.

SECTION 4

Call-Out Boxes

1. Greek
2. kings, priests, and other important people
3. Untouched by raiders, we can learn about Egyptian art and burial practices from it.

Challenge Activity

Students' messages should reflect an understanding of how Egyptian hieroglyphics represented sounds.

History of Ancient Kush

COMPREHENSION AND CRITICAL THINKING

1. Answers may include pyramids, writing systems, houses, and religious symbols.
2. Kushite women were active members of society; Egyptian women were expected to remain within the home.
3. possible answer—Resources/cultures were different, so the architecture was different.
4. possible answers—Similar: both were advanced, involved writing, and worshipped lion-like beings; Different: the societal roles of women, architectural designs, and forms of writing

SECTION 1

Call-Out Boxes

1. gold, copper, and stone
2. around 2000 BC
3. about 450 years
4. iron; bronze

Challenge Activity

possible answer—A leader's desire for power may be so strong that he or she is willing to do anything to acquire it.

SECTION 2

Call-Out Boxes

1. the iron industry
2. north
3. Women had more opportunity to achieve success in Kush.
4. King Ezana

Challenge Activity

Students' essays should include the steps to reach their goal of a rich and powerful Kush.

History of West Africa

COMPREHENSION AND CRITICAL THINKING

1. Farmers banded together to defend themselves from invading herders. They grew more powerful and used iron to make weapons and farming tools.
2. Ghana controlled the West African trade routes, collected taxes, and increased its size by conquering other tribes. Ghana's decline was caused by three factors: Muslim Berbers invaded; herded animals overgrazed and ruined the farmland; internal rebellions led to loss of order.
3. possible answers—Trade routes existed to fill a need. If one empire declines, the need for the trade would not be lessened. The most powerful remaining group would gain power by controlling trade routes, and a new empire would form.
4. Moroccan invaders destroyed Songhai's centers of power, trade declined, and the empire never recovered.

SECTION 1

Call-Out Boxes

1. the controlling of trade routes
2. The fact that traders wanted to keep the location of the gold mines secret shows that it was more valuable than salt.
3. Gold became more valuable because only a few people owned it.
4. Muslim Berbers invaded, animals overgrazed the farmland, internal rebellions caused Ghana to lose order

Challenge Activity

possible advantages—secrecy about sources and easier movement of materials; possible disadvantages—costs of "middlemen" and slower progress in negotiations

SECTION 2

Call-Out Boxes

1. Niger River
2. He captured important trading cities, brought attention to Mali on his pilgrimage to Mecca, and spread Islam by building mosques.
3. Sunni Baru was not Muslim. People rebelled, fearing the loss of trade power if he did not support Islam.
4. Most sources estimate the population of Timbuktu today as being between 15,000–30,000.

Challenge Activity

Answers will vary and may mention the king's wealth, great learning, and support for Islam.

SECTION 3

Call-Out Boxes

1. No; although West Africans used Arabic in their writing, they passed along their oral history in their native languages.
2. Names and events were not always told accurately, and because the stories were passed on orally, there are no written records to prove their accuracy.
3. Henri Matisse and Pablo Picasso
4. entertainment, celebration of history, religious celebrations

Challenge Activity

possible answers—Oral histories are less likely to be biased against the culture and may be more accurate than histories conveyed by outsiders who might miss some of the subtle meanings of the culture; written history is more accurate because it is easier to replicate exactly. Students should correctly describe the primary and secondary sources.

North Africa

COMPREHENSION AND CRITICAL THINKING

1. Algeria: Algiers; Tunisia: Tunis; Libya: Tripoli; Egypt: Cairo
2. They provide water, food, grazing land for livestock, and an area to rest.
3. Egyptian farmers grow crops on irrigated land along the Nile near oases. Dry farming is pursued along the northern coast of the countries west of Egypt.
4. possible answer—Islam because most North Africans are Muslim, and issues concerning the role of Islam are matters of serious debate in the region.

SECTION 1

Call-Out Boxes

1. Egypt, Libya, Tunisia, Algeria, Morocco
2. runs the length of Egypt, becomes a delta in the north, empties in the Mediterranean, irrigates the farmland along its banks
3. It provides water, food, shade, and a place to rest.
4. a desert climate

Challenge Activity

The fertile land along the Nile supports farming. The harsh desert conditions of most of Egypt make living elsewhere difficult.

SECTION 2

Call-Out Boxes

1. in pyramids
2. Egyptian pictures and symbols that stand for ideas and words
3. from the AD 600s to the 1800s
4. Spain, France, Italy, and Britain
5. Arabic; Islam
6. Muhammad's birthday and Ramadan

Challenge Activity

Answers will vary but should be supported with details and information from the text.

SECTION 3

Call-Out Boxes

1. overcrowding, traffic, pollution
2. battling poverty, illiteracy, disease, and pollution; disagreement over the role of Islam in society
3. Most of North Africa's trading partners are members of the European Union.
4. oil production

Challenge Activity

possible answer—North African countries must balance Muslim religious influences and the laws that govern each country with personal freedoms, economic realities, and international relationships.

West Africa

COMPREHENSION AND CRITICAL THINKING

1. Culture
2. Climate and land make it very hard to farm and feed their people.
3. Atlantic slave trade, European colonies in Africa
4. Corrupt government and internal conflict such as civil wars have kept West African economies from developing.

SECTION 1

Call-Out Boxes

1. It brings water to the people of the region for farming and fishing. It also provides a transportation route.
2. the Sahara
3. overgrazing, trees cut for firewood
4. to make room for the growing population
5. diamonds, gold, iron ore, and bauxite

Challenge Activity

Because of the desert climate, people cannot farm or raise animals; there is little or no freshwater available in the northern region.

SECTION 2

Call-Out Boxes

1. It had a university, mosques, and many schools.
2. 1600s; 1800s
3. African cultures, European culture, and Islam
4. Islam, Christianity, animism
5. There are few trees for wood.

Challenge Activity

possible answers—Both have diverse ethnic groups; many religions; many languages, although most people in the U.S. speak English as well as their native language.

SECTION 3

Call-Out Boxes

1. Abuja
2. high birthrate, corrupt government
3. democratic
4. Ghana and Côte d'Ivoire
5. drought and locusts

:hallenge Activity

Jigeria—rich in resources, strong economy, ut imports food, and its people are very poor; il is an important export for both Nigeria nd Chad; most West African countries are oor with corrupt or unstable governments.

:ast Africa

:OMPREHENSION AND CRITICAL 'HINKING

1. ethnic conflicts
2. details could be added to each column
3. terrorist attacks

;ECTION 1

:all-Out Boxes

1. Although it's near the equator, its peak is covered with ice and snow.
2. elephants, giraffes, lions, and zebras
3. the Nile
4. Lake Victoria
5. crops fail, cattle die, and people begin to starve
6. They receive a lot of rainfall and have a mild climate.

:hallenge Activity

tudents' booklets should include details about ıe landforms, rivers, and lakes in East Africa.

ECTION 2

'all-Out Boxes

1. Ethiopia and Nubia
2. Muslim Arabs from Egypt
3. They used imperialism.
4. Ethiopia
5. French, English, and Swahili

hallenge Activity

tudents should incorporate information bout the former colonial rulers and the hallenges facing the newly independent ɔuntry today.

ECTION 3

all-Out Boxes

1. Tanzania—Dodoma; Kenya—Nairobi
2. the Tutsi and Hutu
3. Muslims and Christians
4. Ethiopia, Eritrea, Somalia, and Djibouti
5. tourism and farming

Challenge Activity

Students' proposals should offer solutions to problems such as drought, famine, civil war, and economic improvement.

Central Africa

COMPREHENSION AND CRITICAL THINKING

1. different languages and customs
2. It is rich in resources, but it also makes transportation of goods difficult which is an obstacle to economic growth.
3. They have kept Central African economies from developing.

SECTION 1

Call-Out Boxes

1. east of the basin along the Western Rift Valley
2. led to the growth of dense tropical forests
3. possible answers—forests, rivers, oil, natural gas, coal, minerals
4. in the copper belt

Challenge Activity

possible answer—The Congo River provides a transportation route for carrying minerals to the ports on the Atlantic Ocean.

SECTION 2

Call-Out Boxes

1. from trade in animal skins, shells, slaves, and ivory
2. The Europeans first came looking for trade goods.
3. It made some African kingdoms rich, but millions of enslaved Africans were sent to colonies in the Americas.
4. Colonial borders put groups of people together who had different languages and customs, which led to conflicts.
5. During the Cold War, both countries supported different groups in small wars in Africa. These wars killed many people.
6. sculpture, carved wooden masks, cotton gowns, *likembe*, and *makossa*

Challenge Activity
possible answers—Hurt: imposed boundaries on the people of different homelands, encouraged the slave trade, and led to wars after independence; Helped: brought technology and trade to help economies grow

SECTION 3
Call-Out Boxes
1. colonial rule and civil wars
2. civil war, bad government, crime
3. subsistence farming
4. better railroads and ports to help ship resources
5. possible answer—Disease kills many people in Central Africa and results in a young and less-skilled population.

Challenge Activity
possible answer—With rapid population growth, governments may not be able to provide services for the people. A poor economy will not be able to provide food and goods for a large population. The growing population could make governments more unstable and stress the economy.

Southern Africa

COMPREHENSION AND CRITICAL THINKING
1. farming, ranching
2. manufacturing, fishing
3. possible answer—Mining/minerals, because most countries in the region have rich supplies of minerals.

SECTION 1
Call-Out Boxes
1. the steep face at the edge of a plateau or other raised area
2. the Okavango and the Orange rivers
3. low, flat areas into which ancient streams drained and later evaporated
4. gold, diamonds, platinum, copper, uranium, coal, and iron ore

Challenge Activity
Answers will vary but should be supported with details from the section.

SECTION 2
Call-Out Boxes
1. They practiced different religions, languages and customs, and displaced people, which led to conflict between groups; slave trade; increasing racism
2. the Shona
3. Afrikaner frontier farmers
4. The diamond and gold trade was more profitable.
5. the separate areas where blacks had to live under apartheid

Challenge Activity
Possible answers might focus on Britain's cultural influence over languages spoken, religious practices, and democratic governments.

SECTION 3
Call-Out Boxes
1. penalties to force a change in policy
2. the African National Congress, a group defending the rights of black South Africans
3. Nelson Mandela
4. Botswana is rich in mineral resources wit[h] cattle ranching and diamond mining as i[ts] main economic activities. Zimbabwe has [a] poor economy.
5. taxes collected on products shipped out o[f] its ports
6. unique plants and animals
7. The African Union promotes cooperation among African countries.

Challenge Activity
Possible responses might focus on the region's mineral resources, which could be distributed more fairly to help reduce poverty and disease.